Activities Manual for

Programmable Logic Controllers

Frank D. Petruzella

Activities Manual for

Programmable
Logic
Controllers

Fourth Edition

McGraw Hill

*Connect
Learn
Succeed*™

The McGraw·Hill Companies

Mc Graw Hill — Connect Learn Succeed™

Activities Manual to accompany
Programmable Logic Controllers, Fourth Edition
Frank D. Petruzella

Published by McGraw-Hill, a business unit of The McGraw-Hill Companies, Inc., 1221 Avenue of the Americas, New York, NY 10020. Copyright © 2011 by The McGraw-Hill Companies, Inc.
All rights reserved. Previous editions © 1989, 1998, and 2005.

2 3 4 5 6 7 8 9 0 QDB/QDB 1 0 9 8 7 6 5 4 3 2 1

ISBN   978-0-07-330342-0
MHID   0-07-330342-9

www.mhhe.com

Contents

Preface

This *Activities Manual to accompany Programmable Logic Controllers, 4/e* provides a wide variety of test questions and programming assignments designed to aid students in understanding the underlying principles of programmable logic controllers.

The *Activities Manual* parallels the text chapters in its treatment of subject material. Each chapter in this manual contains multiple-choice, completion, and true/false test questions. These questions are designed primarily to measure the student's knowledge of the material presented in the text. The on-line **Instructor's Resource Center** contains the answers to all *Activities Manual* test questions.

The *Activities Manual* programming assignments are designed to provide hands-on programming tasks tied to the text chapters. The range and the type of programming scenarios covered here will enable students to gain practical experience and valuable insights into the most current PLC automation technology. All programming assignments are generic in nature which allows them to be implemented using PLCs from different manufactures. Programs will vary to some degree with the particular PLC used and so no answers for these assignments have been provided.

I would like to thank Brent Garner, McNeese State University, for his input and review of the Activities Manual.

Frank D. Petruzella

About the Author

Frank D. Petruzella has extensive practical experience in the electrical control field, as well as many years of experience teaching and authoring textbooks. Before becoming a full time educator, he was employed as an apprentice and electrician in areas of electrical installation and maintenance. He holds a Master of Science degree from Niagara University, a Bachelor of Science degree from the State University of New York College–Buffalo, as well as diplomas in Electrical Power and Electronics from the Erie County Technical Institute.

CHAPTER 1

Programmable Logic Controllers (PLCs)
An Overview

TEST 1.1

Choose the letter that best completes the statement. Place the answers in the column at the right.

1. PLCs were originally designed as replacements for   1._____
a) microcomputers.   c) analog controllers.
b) relay control panels.   d) digital controllers.

2. Basically, the function of a PLC is to   2._____
a) amplify various weak signal sources.
b) control a high voltage output with a low voltage input.
c) control the speed of motors.
d) make logical decisions and control outputs based on them.

3. Modifying relay-type process control circuits usually involves changing   3._____
the
a) circuit wiring.   c) output circuit modules.
b) input circuit modules.   d) circuit operating voltage levels.

4. Which of the following is *not* an advantage that PLCs offer over the   4._____
conventional relay-type of control system?
a) Smaller size   c) Higher current capacity
b) Less expensive   d) More reliable

5. The main difference between a PLC and relay control system is that   5._____
a) different types of input devices are used.
b) different types of output devices are used.
c) different input and output voltage levels are used.
d) one uses hardwired relay control logic and the other uses programmed instructions.

6. The central processing unit   6._____

a) looks at the inputs, makes the decisions based on the program, and sets the outputs.

b) looks at the outputs, makes the decisions based on the program, and sets the inputs.

c) serves only to store the program in memory.

d) serves only to supply power to the backplane.

7. PLC proprietary architecture   7._____

a) is the opposite to open architecture.

b) makes it more difficult to connect to devices made by other PLC manufacturers.

c) does not allow programs to be interchanged between different PLC manufacturers.

d) all of the above

8. The output interface module connects to   8._____

a) sensing devices such as switches or pushbuttons.

b) load devices such as lamps or solenoids.

c) a programming device such as a computer.

d) all of the above

9. Field or real-world devices refer to   9._____

a) input devices only.

b) output devices only.

c) load devices only.

d) all devices that are physically wired to the PLC.

10. The power required to operate the logic circuits of the processor unit   10. _____
is typically

a) low voltage AC.   c) low voltage DC.

b) high voltage AC.   d) high voltage DC.

11. The control plan stored in the PLC is called   11._____

a) a program.   c) FORTRAN.

b) a Boolean ladder.   d) a microprocessor.

2

12. The programming device   12._____

a) is used to enter the program into the memory of the processor.

b) is commonly a personal computer.

c) can be a handheld device.

d) all of the above

13. The programming device must be connected to the controller   13._____

a) at all times.   c) when monitoring a program.

b) when entering a program.   d) both b and c.

14. The   ⊣├ symbol in a ladder logic diagram   14._____

a) can be thought of as a normally open contact.

b) represents a capacitor.

c) is always at logic 0.

d) is always at logic 1.

15. The   ⊣( )├ symbol in a ladder logic diagram represents a   15._____

a) set of normally closed contacts.   c) seal-in contact.

b) virtual relay coil.   d) field input sensing device.

16. When a field device contact connected to the input module closes   16._____

a) a logic 1 is recorded in the memory location of the coil with the same address.

b) a logic 1 is recorded in the memory location of the contact with the same address.

c) a logic 0 is recorded in the memory location of the coil with the same address.

d) a logic 1 is recorded in the memory location of the contact with the same address.

17. At the start of the PLC scan the   17._____

a) status of all inputs is read.

b) status of all outputs is updated.

c) program is executed.

d) diagnostics and communications tasks are executed.

18. The scan time is the time required   18._____

a) to record the status of all input devices.

b) to record the status of all output devices.

c) to execute one cycle of the total program.

d) for the information to pass from input to output.

19. Unlike personal computers, PLCs are
a) equipped with input and output modules.
b) equipped with a control programming language.
c) designed for the industrial environment.
d) all of the above

19._____

20. A human machine interface (HMI)
a) allows the user to monitor a process.
b) allows the user to control a process.
c) can provide a graphical representation of a process.
d) all of the above

20._____

21. Programmable logic controllers are categorized according to the
a) number of I/O points.
b) current rating of I/O modules.
c) power rating of the I/O modules.
d) cost of the I/O modules.

21._____

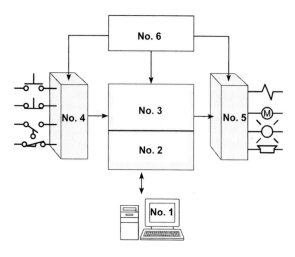

Figure 1-1 Block diagram for question 22.

22-1. In the PLC block diagram of Figure 1-1, block No. 1 represents the
a) CPU unit.
b) programming device.
c) input module.
d) output module.

22-1._____

22-2. Block No. 2 represents the
a) memory.
b) programming device.
c) input module.
d) power supply module.

22-2._____

22-3. Block No. 3 represents the

22-3.____

a) CPU unit.

c) input module.

b) programming device.

d) output module.

22-4. Block No. 4 represents the

22-4.____

a) memory.

c) input module.

b) programming device.

d) CPU.

22-5. Block No. 5 represents the

22-5.____

a) memory.

c) input module.

b) power supply module.

d) output module.

22-6. Block No. 6 represents the

22-6.____

a) processor module.

c) input module.

b) power supply module.

d) output module.

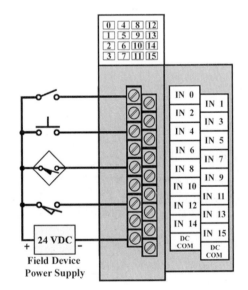

Figure 1-2 Diagram for question 23.

23.1. The diagram of Figure 1-2 is that of a(n)

23-1.____

a) relay schematic.

c) input module wiring.

b) ladder logic program.

d) output module wiring.

23-2. The voltage that would be present between the DC common and

23-2.____

terminal 4 with the pushbutton open would be approximately

a) 0 volts.

c) 12 volts.

b) 6 volts.

d) 24 volts.

23-3. The voltage that would be present between the DC common and 23-3.____
terminal 4 with the pushbutton closed would be approximately
a) 0 volts. c) 12 volts.
b) 6 volts. d) 24 volts.

23-4. The devices connected to the terminals would be classified as 23-4.____
a) field input devices. c) field output devices.
b) internal input instructions. d) internal output instructions.

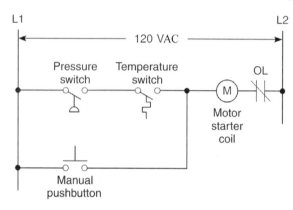

Figure 1-3 Diagram for question 24.

24-1. The diagram of Figure 1-3 is that of a(n) 24-1.____
a) hardwired relay schematic. c) input module schematic.
b) ladder logic program. d) output module schematic.

24-2. In order to energize the starter coil 24-2.____
a) the pressure switch, and the temperature switch, and the manual
pushbutton must be closed.
b) the pressure switch, or the temperature switch, or the manual pushbutton
must be closed.
c) the pressure switch, and the temperature switch, or the manual pushbutton
must be closed.
d) all of the above

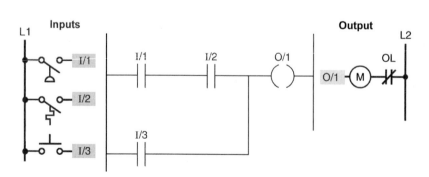

Figure 1-4 Diagram for question 25.

25-1. The diagram of Figure 1-4 is that of a(n)   25-1.____

a) relay schematic.   c) input module wiring diagram.

b) ladder logic program.   d) output module wiring diagram.

25-2. For there to be a continuous logic path from left to right across the   25-2.____

rung,

a) I/1, I/2, and I/3 must all be at logic 1.

b) I/1, I/2, and I/3 must all be at logic 0.

c) I/1 and I/2 or I/3 must be at logic 1.

d) I/1 and I/2 or I/3 will be at logic 0.

26. The PLC power supply module normally is rated to provide the   26._____

power for

a) all field devices.   c) output field devices only.

b) input field devices only.   d) PLC backplane and I/O modules.

27. Which module of the PLC is responsible for performing logical   27._____

operations?

a) Processor   c) Output

b) Input   d) Power supply

28. Which module of the PLC connects directly to field devices such as pilot   28._____

lights, motor starters, and solenoids?

a) Input   c) Power supply

b) Output   d) Memory

29. _____ I/Os are typical of small PLCs that come in one package with no   29._____

separate removable units.

a) Fixed   c) Digital

b) Modular   d) Analog

30. PLC software that runs on a personal computer can be used to   30._____

a) write a PLC program.   c) monitor the control process.

b) document a PLC program.   d) all of the above

31. A control management PLC application normally requires a   31._____
a) micro-size PLC.   c) medium-size PLC.
b) small-size PLC.   d) large-size PLC.

32. Which of the following is *not a* factor affecting the memory size needed   32._____
for a particular PLC installation?
a) Voltage rating of field devices   c) Size of control program
b) Number of I/O points   d) Supervisory functions required

TEST 1.2

Place the answers in the column at the right.

1. Programmable logic controllers were originally designed to perform logic   1._____
functions previously accomplished by _____.

2. The number and type of I/Os cannot be changed in a fixed PLC.   2._____
(True or False)

3. In a PLC system, there is a physical connection between field input   3._____
devices and output devices. (True or False)

4. Identify the following electrical components by specifying whether they are input field
devices or output field devices.
a) Pushbutton   4a._____
b) Solenoid   4b._____
c) Pilot lamp   4c._____
d) Selector switch   4d._____

5. In a typical ladder logic program the symbols represent the (a) __ and the   5a._____
numbers represent the (b) ___.   5b._____

6. The scan time is the time required for one complete execution of the   6._____
user program. (True or False)

7. The input/output system forms the interfaces through which field devices   7._____
are connected to the controller. (True or False)

8. ____ is the process of reading inputs, executing the program, and setting   8. _____
outputs on a continuous basis.

9. The abbreviation I/O means (a) ___ and (b) ___.   9a._____
   9b._____

10. Plug-in compartments allow I/O modules to be easily connected   10._____
and replaced. (True or False)

11. To operate the program, the controller is placed in the ___ mode.

11._____

12. If there is no continuous logic path from left to right on the program rung, the output coil status is set to ___.

12._____

13. Changes to hardwired relay control systems usually require some ___ of the system.

13._____

14. A personal computer communicates with the PLC processor via a serial or parallel data communications link. (True or False)

14._____

15. The programming device must be connected to the controller to run the program. (True or False)

15._____

16. Incoming control signals to a PLC are called _____.

16._____

17. Signals going out from a PLC to control field devices are called _____.

17._____

18. PLC systems usually require as much space in an enclosure as equivalent hardwired relay systems. (True or False)

18._____

19. The term *central processing unit is* often used interchangeably with the term ___.

19._____

20. What is the name of the most common programming language used in PLCs?

20._____

21. The PLC program is stored in the processor module's ___.

21._____

22. A PLC is basically a computer designed for use in electrical control applications. (True or False)

22._____

23. The programmable controller operates in real time. (True or False)

23._____

24. When a module is slid into a PLC rack, it makes electrical connection with the _____.

24._____

25. One disadvantage of modular I/O is its lack of flexibility. (True or False)

25._____

26. A PLC power supply module does not normally supply power to the field devices. (True or False)

26._____

27. Removing the programming device from the PLC will not affect the operation of the user program. (True or False)

27._____

28. Software installed and run on a personal computer can be used to write a PLC program. (True or False)

28._____

29. The instruction set for a particular PLC lists the types of instructions supported. (True or False)

29._____

30. When dealing with PLC memory, one K of memory represents 1024. (True or False)

30._____

31. The number of I/O points does not affect the memory size required for a PLC installation. (True or False)

31._____

Programming Assignments

1a) On a separate sheet of paper, draw an I/O wiring diagram and ladder logic program for the relay schematic shown in Figure 1-5. Use the field devices with the appropriate addressing for the PLC trainer you will be working with.

   b) Enter the program into the PLC, and prove its operation.

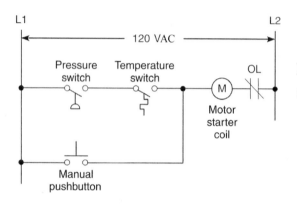

Figure 1-5 Relay schematic for assignment 1.

2a) On a separate sheet of paper, draw a ladder logic diagram for the modified relay ladder schematic shown in Figure 1-6.

   b) Enter the program into the PLC, and prove its operation.

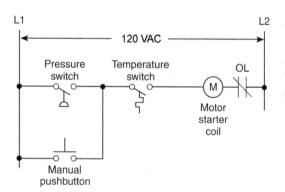

Figure 1-6 Relay schematic for assignment 2.

3a) On a separate sheet of paper, draw a ladder logic program of the relay schematic circuit altered so that the manual pushbutton, pressure switch, and temperature switch all must be closed to permit operation of the motor.

   b) Enter the program into the PLC, and prove its operation.

4a) On a separate sheet of paper, draw a modified ladder logic program for the relay schematic circuit altered so that the motor will operate when either the manual pushbutton, pressure switch, or temperature switch is closed.

  b) Enter the program into the PLC, and prove its operation.

CHAPTER 2

PLC Hardware Components

TEST 2.1

Choose the letter that best completes the statement.

1. A _____ is an example of a device that could be used to provide a discrete input to a PLC.

    1._____

a) pushbutton   c) limit switch

b) selector switch   d) all of the above

2. A _____ is an example of an actuator that could be controlled by a discrete output from a PLC.

    2._____

a) pushbutton   b) motor starter

c) limit switch   d) all of the above

3. A(n) _____ input or output is a continuously variable signal within a designated range.

    3._____

a) discrete   c) BCD

b) digital   d) analog

4. One function of a PLC input interface module is to

    4._____

a) accept signals from field devices and convert them into signals that can be used by the processor.

b) convert signals from the processing unit into values that can be used to control the machine or process.

c) input signals from the programming device and convert them into signals that can be used by the CPU.

d) interpret and execute the user program that controls the machine or process.

5. The location of a specific input or output field device is identified by the processor by means of its

    5._____

a) voltage rating.   c) wattage rating.

b) current rating.   d) address.

6. A discrete output interface module is designed to provide   6._____
a) output voltages only in the 5-VDC range.
b) varying AC or DC voltages depending on the type of module selected.
c) ON/OFF switching of the output field device.
d) binary-coded outputs.

7. The following statement that does *not* apply to the optical isolator   7._____
circuit used in I/O modules is that it
a) separates high voltage and low voltage circuits
b) rectifies AC signals.
c) prevents damage caused by line voltage transients.
d) reduces the effect of electrical noise.

8. Individual outputs of a typical AC output interface module usually have   8._____
a maximum current rating of about
a) 1 A or 2 A.   c) 50 mA or 100 mA.
b) 25 A or 50 A.   d) 250 μA or 500 μA

9. Which of the following input field devices would most likely be used   9._____
with an analog interface input module?
a) Pushbutton   c) Selector switch
b) Limit switch   d) Thermocouple

10. The "ON state input voltage range" specification refers to   10._____
a) the type of voltage device that will be accepted by the input.
b) range of leakage voltage present at the input in its ON state.
c) minimum and maximum output operating voltages.
d) voltage at which the input signal is recognized as being ON.

11. Volatile memory elements can be classified as those that   11._____
a) do not retain stored information when the power is removed.
b) retain stored information when the power is removed.
c) do not require a battery backup.
d) both b and c.

12. _____ memory is used by the PLC's operating system.   12._____

a) RAM   c) Flash

b) EEPROM   d) ROM

13. _____ is a type of memory commonly used for temporary storage of   13._____
data that may need to be quickly changed.

a) RAM   c) EPROM

b) ROM   d) EEPROM

14. The most common form of memory used to store, back up, or transfer   14._____
PLC programs is

a) RAM.   c) EEPROM.

b) Flash EEPROM.   d both b and c.

15. In event of a power interruption, a(n) _____ is used in some processors   15._____
to provide power to the RAM.

a) inductor   c) transistor

b) capacitor   d) resistor

16. Which of the following is *not a* function of a PLC programming   16._____
device?

a) To enter the user program   c) To execute the user program

b) To change the user program   d) To monitor the user program

17. Status indicators are provided on each output of an output module   17._____
to indicate that the

a) load has been operated.

b) input associated with the output is active.

c) module fuse has blown.

d) output is active.

18. The I/O system provides an interface between   18._____

a) input modules and output modules.

b) the CPU and field equipment.

c) the CPU and I/O rack.

d) the I/O rack and I/O modules.

19. The PLC chassis comes in different sizes according to the   19._____

a) size of the program.   c) number of slots it contains.

b) type of I/O modules used.   d) all of the above.

20. The Allen-Bradley SLC-500 address I:2/4 refers to an   20._____

a) input module in slot 4, terminal 2.

b) output module in slot 4, terminal 2.

c) input module in slot 2, terminal 4.

d) output module in slot 2, terminal 4.

21. The Allen-Bradley SLC-500 address O:3/0 refers to an   21._____

a) input module in slot 3, terminal 0.

b) output module in slot 3, terminal 0.

c) input module in slot 0, terminal 3.

d) output module in slot 0, terminal 3.

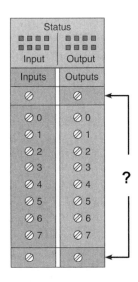

Figure 2-1 I/O module for question 22.

22. For the I/O module of Figure 2-1, the arrows point to the   22._____

a) status indicator connections.   c) output connections.

b) input connections.   d) power supply connections.

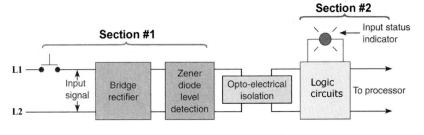

Figure 2-2 Block diagram for question 23.

23. For the block diagram of the input module shown in Figure 2-2,   23._____
Section #1 represents the _____ and #2 the _____.
a) AC, DC   c) power, logic
b) DC, AC   d) logic, power

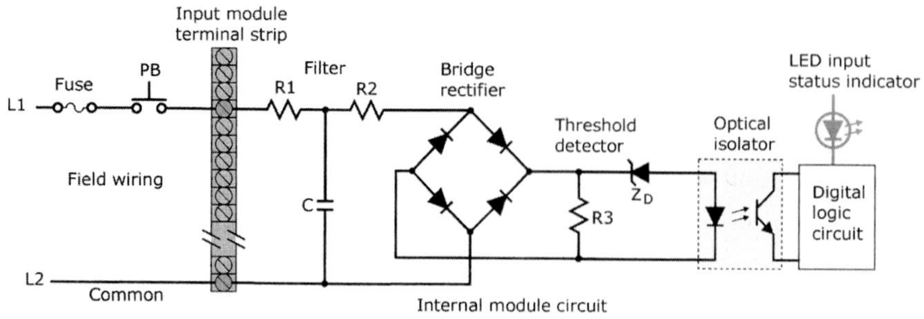

Figure 2-3 Schematic diagram for question 24.

24-1. The schematic diagram of Figure 2-3 is that of a(n)   24-1._____
a) discrete output module.   c) discrete input module.
b) analog output module.   d) analog input module.

24-2. The purpose of the filter section is to   24-2._____
a) aid in fault diagnosis.
b) set the minimum level of voltage that can be detected.
c) protect against electrical noise interference.
d) separate the higher line voltage from the logic circuits.

24-3 The purpose of the zener diode (Z_D) is to   24-3._____
a) aid in fault diagnosis.
b) set the minimum level of voltage that can be detected.
c) protect against electrical noise interference.
d) separate the higher line voltage from the logic circuits.

24-4 The purpose of the LED indicator is to   24-4._____
a) aid in fault diagnosis.
b) set the minimum level of voltage that can be detected.
c) protect against electrical noise interference.
d) separate the higher line voltage from the logic circuits.

24-5 The purpose of the optical isolator is to   24-5.____

a) aid in fault diagnosis.

b) set the minimum level of voltage that can be detected.

c) protect against electrical noise interference.

d) separate the higher line voltage from the logic circuits.

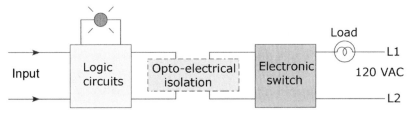

Figure 2-4 Block diagram for question 25.

25. For the block diagram of the output module shown in Figure 2-4,   25._____

the input comes from the

a) input field device.   b) processor.

c) output field device.   d) line power supply.

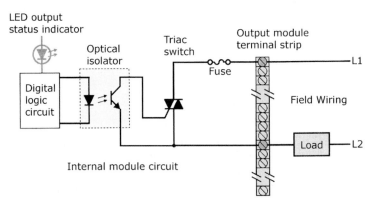

Figure 2-5 Schematic diagram for question 26.

26-1. The schematic diagram of Figure 2-5 is that of a(n)   26-1.____

a) discrete output module.   c) discrete input module.

b) analog output module.   d) analog input module.

26-2. The input signal to the module comes from   26-2.____

a) the input field device.

b) the output field device.

c) internal logic circuitry of the processor.

d) either a or b

26-3 The purpose of the triac switch is to

26-3.____

a) turn the load ON and OFF.

b) vary the current flow to the load in accordance with the input signal level.

c) vary the voltage across the load in accordance with the input signal level.

d) both b and c.

26-4 When the triac is in the OFF state

26-4.____

a) zero current always flows through the load.

b) a small leakage current may flow through the load.

c) the rated surge current flows through the lamp.

d) the rated nominal current flows through the lamp.

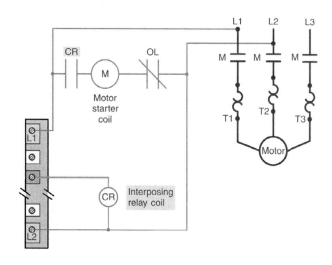

Figure 2-6 Schematic diagram for question 27.

27. The schematic diagram of Figure 2-6 is an example of how a PLC output module is connected to

27.____

a) isolate the load from the controller.   c) vary the speed of a motor.

b) control a high resistance.   d) control a high current load.

28. Which of the following devices can be used for switching the output of a discrete DC output module?

28. _____

a) Transistor   c) Relay

b) Triac   d) Either a or c

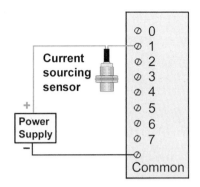

Figure 2-7 Current sourcing sensor for question 29.

29. The current sourcing sensor shown in Figure 2-7 must be matched 29._____
with a _____ PLC input module.
a) current sinking   c) alternating current
b) current sourcing   d) either a or b

30. Typical analog inputs and outputs can vary from   30._____
a) 0 to 20 mA.   c) 0 to 10 volts.
b) 4 to 20 mA.   d) all of the above.

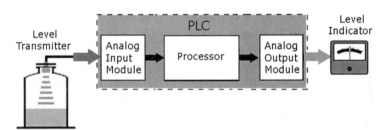

Figure 2-8 Block diagram for question 31.

31. For the block diagram of the analog PLC control shown in   31._____
Figure 2-8, which part has a binary input and analog output value?
a) Level transmitter   c) Processor
b) Input module   d) Output module

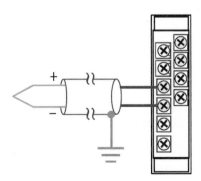

Figure 2-9 Block thermocouple input module for question 32.

32-1. For the thermocouple analog input module shown in Figure 2-9, shielded cable is used to

32-1._____

a) reduce unwanted electrical noise signals.
b) carry the higher current required.
c) lower the resistance of the conductors.
d) insulate the circuit from other cables.

32-2. The thermocouple shown is a(n)

32-2._____

a) ungrounded type with the shield grounded at the module end.
b) ungrounded type with the shield grounded at the thermocouple end.
c) grounded type with the shield grounded at the module end.
d) grounded type with the shield grounded at the thermocouple end.

33. The main element of an analog output module is:

33._____

a) AC to DC rectifier.
b) DC to AC inverter.
c) analog to digital converter.
d) digital to analog converter.

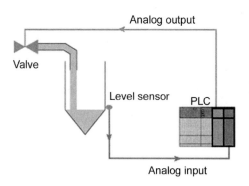

Figure 2-10 Analog I/O system for question 34.

34. For the PLC analog I/O control system shown in Figure 2-10, the fluid flow is controlled by

34._____

a) varying the amount of the valve opening.
b) switching the valve ON and OFF.
c) switching the level sensor ON and OFF.
d) varying the position of the level sensor.

35. Which of the following special I/O modules would be used to operate a seven-segment LED display?

35._____

a) Encoder-counter module
b) BCD-output module
c) Stepper-motor module
d) High-speed counter module

36. A _____ module is used to establish connections for the exchange of data.

36._____

a) thumbwheel

c) servo

b) communication

d) PID

37. High-density I/O modules

37._____

a) may have up to 64 inputs or outputs per module.

b) require more space.

c) can handle greater amounts of current per output.

d) all of the above

38. Discrete I/O modules can be classified as

38._____

a) bit oriented.

c) processor oriented.

b) word oriented.

d) power supply oriented.

39. Which of the following specifications defines the number of field inputs or outputs that can be connected to a single module?

39._____

a) Electrical isolation

c) Threshold voltage

b) Points per module

d) Current per input

40. The _____ of an analog I/O module specifies how accurately an analog value can be represented digitally.

40._____

a) number of inputs and outputs per card

b) input impedances and capacitances

c) resolution

d) common mode rejection ratio

41. The processor module of the PLC is where the:

41._____

a) ladder logic program is stored.

b) input connections are made.

c) output connections are made.

d) sensors are located.

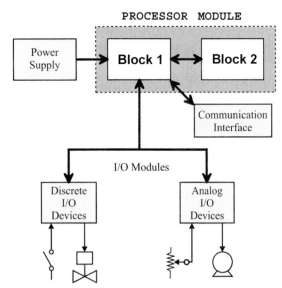

PROCESSOR MODULE

Power Supply → Block 1 ↔ Block 2

Communication Interface

I/O Modules

Discrete I/O Devices

Analog I/O Devices

Figure 2-11 Processor module for question 42.

42. For the processor module shown in Figure 2-11, Block 1 represents the ___ and Block 2 the ___.

42._____

a) input, output
b) output, input
c) memory, CPU
d) CPU, memory

43. When placed in the _____ mode, the processor does not scan/execute the ladder program.

43._____

a) program
b) run
c) test
d) remote

44. The most commonly used programming device is a

44._____

a) personal computer.
b) dedicated industrial programming terminal.
c) handheld programmer.
d) proprietary programming device.

45. Electronic components found in PLC modules

45._____

a) are not affected by electrostatic voltages.
b) can be damaged by electrostatic voltages.
c) can have their performance degraded by electrostatic voltages.
d) both b and c

46. Batteries are used in a PLC's processor to					46._____

a) operate the status light LEDs.

b) maintain data in volatile memory when line power is removed from the processor.

c) maintain data in nonvolatile memory when line power is removed from the processor.

d) maintain outputs through a power failure.

TEST 2.2

Place the answers to the following questions in the answer column at the right.

1. An analog input or output is a signal that varies continuously within a certain range. (True or False)

1._____

2. The I/O section of a PLC system can consist _____ of an I/O rack and individual I/O   .

2._____

3. The location of a module within a rack and the terminal number of a module to which an input or output device is connected will determine the device's ____.

3._____

4. Most input modules have blown fuse indicators. (True or False)

4._____

5. The I/O address is used by the processor to identify where the device is ____.

5._____

6. A standard I/O module consists of a(n) (a) _____ board and a(n) (b) _____ assembly.

6a._____
6b._____

7. I/O modules are designed to plug into a slot or connector. (True or False)

7._____

8. Discrete I/O interfaces allow only _____ type devices to be connected.

8._____

9. I/O modules' circuitry can be divided into two basic sections: the (a) _____ section and the (b) _____ section.

9a._____
9b._____

10. Optical isolation used in I/O modules helps reduce the effects of electrical noise. (True or False)

10._____

11. AC output modules often use a solid-state device such as a(n) _____ to switch the output ON and OFF.

11._____

12. I/O modules are keyed to prevent unauthorized personnel from removing them from the I/O rack. (True or False)

12._____

13. The maximum current rating for the individual outputs of an AC output module is usually in the 20- to 30-ampere range. (True or False)

13._____

14. A(n) _____ relay is used for controlling larger load currents.

14._____

15. Analog input interface modules contain a(n) _____ converter circuit.

15._____

16. A thermocouple would be classified as an analog input sensing device. (True or False)

16._____

17. Shielded twisted pair cable is used for connecting to thermocouple inputs to reduce unwanted electrical noise. (True or False)

17._____

18. Electrical noise usually causes permanent operating errors. (True or False)

18._____

19. Match each of the following specifications with the appropriate description. Place the number from the specifications list in the answer column.

SPECIFICATION
  1) nominal current per input
  2) ON-state input voltage range
  3) OFF-state leakage current
  4) electrical isolation
  5) input delay
  6) nominal input voltage
  7) surge current
  8) output voltage range
  9) maximum output current rating
 10) nominal output voltage

DESCRIPTION

a) Maximum voltage isolation between the I/O circuits and the controller logic circuitry.

19a._____

b) Maximum value of current that flows through the output in its OFF state.

19b._____

c) Maximum inrush current and duration an output module can withstand.

19c._____

d) Maximum current that a single output and the module as a whole can safely carry.

19d._____

e) Minimum and maximum output operating voltages.

19e._____

f) Magnitude and type of voltage source that can be controlled by the output.

19f._____

g) Duration for which the input must be ON before being recognized as a valid input.

19g._____

h) Minimum input current that the input device must be capable of driving to operate the input circuit.

19h._____

i) Voltage level at which the input signal is recognized as being ON.

19i._____

j) Magnitude and type of voltage signal that will be accepted by the input.

19j._____

20. The processor continually interacts with the _____ to interpret and execute the user program.

20._____

21. The processor may perform functions such as timing, counting, and comparing in addition to logic processing. (True or False)

21._____

22. Memory is where the control plan is held or stored in the controller. (True or False)

22._____

23. One ___ is a memory location that may store one binary number that has the value of either 1 or 0.

23._____

24. A volatile memory will lose its programmed contents if operating power is lost. (True or False)

24._____

25. A nonvolatile memory will retain its programmed contents if operating power is lost. (True or False)

25._____

26. RAM memory is nonvolatile. (True or False)

26._____

27. Information stored in a RAM memory location can be written into or read from. (True or False)

27._____

28. When a new program is loaded into a PLC's memory, the old program that was stored in the same locations is overwritten and essentially erased. (True or False)

28._____

29. The type of battery typically used PLC processors is ____.

29._____

30. Flash memory functions similar to ___ memory.

30._____

31. Most PLC programming software will allow you to develop programs on another manufacturer's PLC. (True or False)

31._____

32. Analog signals can have only two states. (True or False)

32._____

33. Memory modules used to copy a program from one PLC to another usually contain ___ memory.

33._____

34. A modular PLC that has room for several I/O modules is capable of being customized for a particular application. (True or False)

34._____

35. Remote I/O racks are linked to the local rack through a(n) ____ module.

35._____

36. In general, rack/slot-based addressing elements include (a) __, (b) __, and (c) __.

36a._____
36b._____
36c._____

37. I/O modules are normally installed or removed while the PLC is powered. (True or False)

37._____

38. A module inserted into the wrong slot could be damaged. (True or False)

38._____

39. Modules receive voltage and current for proper operation from the _____ of the rack enclosure.

39._____

40. The two basic types of analog input modules are (a) ___ sensing and (b) ___ sensing.

40a._____

40b._____

41. Intelligent I/O modules have their own _____ on board.

41._____

42. A redundant PLC system is configured using two processors. (True or False)

42._____

43. Most PLC electronic components are not sensitive to electrostatic discharge. (True or False)

43._____

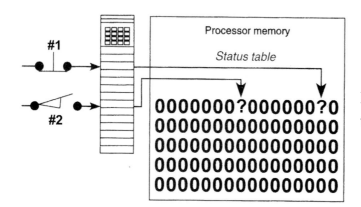

Figure 2-12 I/O module and table for question 44.

44. Answer each of the following for the I/O module and status table shown in Figure 2-12.

a) The type of module shown is a(n) ___ (discrete or analog) module.

44a._____

b) The type of image table shown is a(n) ___ image table.

44b._____

c) The status light indicator associated with device #1 would be ____. (ON or OFF)

44c._____

d) The status light indicator associated with device #2 would be ____. (ON or OFF)

44d._____

e) The value stored in memory for device #1 would be ___.

44e._____

f) The value stored in memory for device #2 would be ___.

44f._____

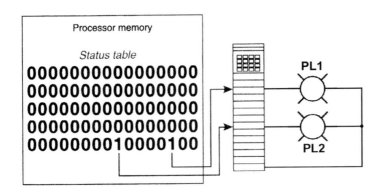

Processor memory

Status table

0000000000000000
0000000000000000
0000000000000000
0000000000000000
0000000010000100

PL1

PL2

Figure 2-13 I/O module and table for question 45.

45. Answer each of the following for the I/O module and status table shown in Figure 2-13.

a) The type of module shown is a(n) __ (discrete or analog) module.   45a._____

b) The type of image table shown is a(n) __ image table.   45b._____

c) The status light indicator associated with PL1 would be ____.   45c._____
(ON or OFF)

d) The status light indicator associated with PL2 would be ____.   45d._____
(ON or OFF)

e) PL1 would be switched ____. (ON or OFF)   45e._____

f) PL2 would be switched ____. (ON or OFF)   45f._____

46. One advantage of discrete relay contact output modules is that they can be used with AC or DC devices. (True or False)   46._____

47. If you had a handheld programming terminal from one manufacturer, you can program only that manufacture's PLC using it. (True or False)   47._____

48. Hot swappable I/O modules are designed to be changed with the power on and the PLC operating. (True or False)   48._____

(a)

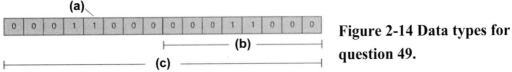

| 0 | 0 | 0 | 1 | 1 | 0 | 0 | 0 | 0 | 0 | 1 | 1 | 0 | 0 | 0 |

(b)

(c)

Figure 2-14 Data types for question 49.

49. Identify data types (a) __, (b) __, and (c) __ shown in Figure 2-14.   49a._____
        49b._____
        49c._____

50. HMI screens are developed using software package on a PC which is   50._____
downloaded into the PLC operator interface device. (True or False)

51. *Discrete* means that each input or output has two states: true (ON) or   51._____
false (OFF). (True or False)

52. Light is used in I/O modules to separate the real-world electrical signals   52._____
from the PLC internal electronic system.(True or False)

53. Digital modules are also called discrete modules. (True or False)   53._____

54. The sum of the backplane current drawn for all modules in a chassis is   54._____
used to select the appropriate chassis power supply rating. (True or False)

Programming Assignments

1) For the PLC you will be working with, summarize the specifications for the
a) input module(s).
c) processor.
b) output module(s).
d) power supply.

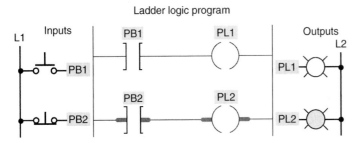

Figure 2-15 Program for assignment 2.

2a) Program your controller to operate according to Figure 2-15.
 b) Download the program to the PLC.
 c) Run the program, and observe the status of the bits stored in the input and output image tables.

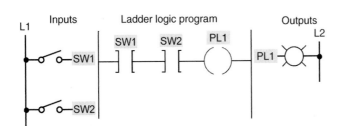

Figure 2-16 Program for assignment 3.

3a) Program your controller to operate according to Figure 2-16.
 b) Download the program to the PLC.
 c) Run the program, and observe the status of the bits stored in the input and output image tables.

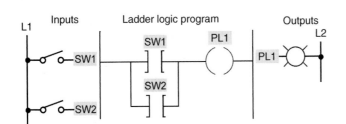

Figure 2-17 Program for assignment 4.

4a) Program your controller to operate according to Figure 2-17.

b) Download the program to the PLC.

c) Run the program, and observe the status of the bits stored in the input and output image tables.

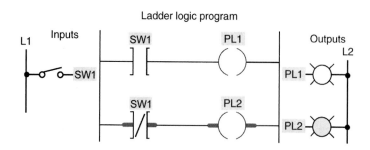

Figure 2-18 Program for assignment 5.

5a) Program your controller to operate according to Figure 2-18.

b) Download the program to the PLC.

c) Run the program, and observe the status of the bits stored in the input and output image tables.

CHAPTER **3**   **Number Systems and Codes**

TEST 3.1

Choose the letter that best completes the statement.

1. The decimal system has as its base   1._____
a) 2.   c) 8.
b) 5.   d) 10.

2. Which of the following number systems has a base of 16?   2._____
a) Hexadecimal   c) Binary-coded decimal
b) Octal   d) Gray code

3. In any number system, the position of a digit that represents part of the   3._____
number has a "weight" associated with its value. The place weights for
binary,
a) start with 1 and are successive powers of 2.
b) increase by adding 2 for each place, starting with 0.
c) increase by adding 2 for each place, starting with 2.
d) start with 2 and double for each successive place.

4. The number 12 is   4._____
a) 12 in any number system.   c) 12 in binary.
b) 12 in decimal.   d) all of the above.

5. The decimal number 15 would be written in binary as   5._____
a) 1111.   c) 4C.
b) 1000.   d) 00011001.

6. The binary number 101 has the decimal equivalent of   6._____
a) 3.   c) 41.
b) 101.   d) 5.

7. The number 127 could *not* be   7._____
a) decimal.   c) octal.
b) hexadecimal.   d) binary.

8. The octal number 153 would be written in binary as   8._____
a) 011 101 001.   c) 011 111 101.
b) 001 101 011.   d) 010 100 011.

9. The binary number 101101 would be written in decimal as   9._____
a) 21.   c) 45.
b) 36.   d) 62.

10. The decimal number 28 would be written in binary as   10._____
a) 11100.   c) 10110.
b) 00111.   d) 01011.

11. The octal number 62 would be written in decimal as   11._____
a) A12.   c) 50.
b) F35.   d) 98.

12. The hexadecimal number C4 would be written in decimal as   12._____
a) 21.   c) 182.
b) 48.   d) 196.

13. The hexadecimal number 2D9 would be written in binary as   13._____
a) 0010 1101 1001.   c) 1100 1111 0010.
b) 1001 1011 0010.   d) 0010 1011 1001.

14. The decimal number 213 would be written in BCD as   14._____
a) 0010 0001 0011.   c) 0111 1001 0011.
b) 1101 1000 1100.   d) 1011 1101 0101.

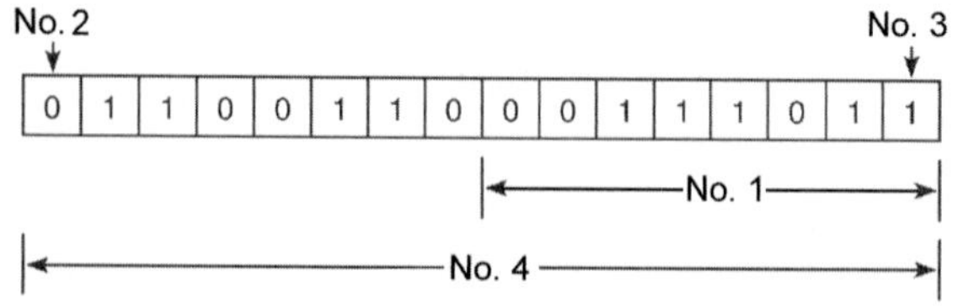

Figure 3-1 Data for question 15.

15-1. One byte of the data shown in Figure 3-1 is represented by

15-1._____

a) No. 1.

c) No. 3.

b) No. 2.

d) No. 4.

15-2. The MSB of the data shown in Figure 3-1 is represented by

15-2._____

a) No. 1.

c) No. 3.

b) No. 2.

d) No. 4.

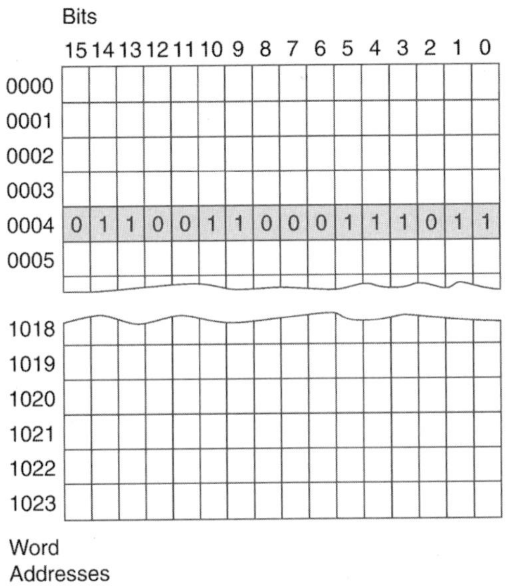

Bits

15 14 13 12 11 10 9 8 7 6 5 4 3 2 1 0

	15	14	13	12	11	10	9	8	7	6	5	4	3	2	1	0
0000																
0001																
0002																
0003																
0004	0	1	1	0	0	1	1	0	0	0	1	1	1	0	1	1
0005																

1018																
1019																
1020																
1021																
1022																
1023																

Word
Addresses

Figure 3-2 Memory size for question 16.

16. Figure 3-2 represents a memory size of

16._____

a) 1023 K.

c) 500 K.

b) 1000 K.

d) 1 K.

17. The main advantage of using the Gray code is

17._____

a) only one digit changes as the number increases.

b) it can be easily converted to decimal numbers.

c) large decimal numbers can be written using fewer digits.

d) it uses the number 2 as its base.

18. The acronym BCD stands for

18._____

a) binary-coded decimal.

b) binary code decoder.

c) base code decoder.

d) base-coded decimal.

19. For a base 8 number system, the *weight value* associated with the third 19._____
digit would be
a) 16.   c) 64.
b) 32.   d) 512.

20. All digital computing devices operate using the binary number system 20._____
because
a) most people are familiar with it.
b) large decimal numbers can be represented in a shorter form.
c) digital circuits can be easily distinguished between two voltage levels.
d) all of the above

21. If a given memory unit consists of 1250 16sixteen-bit words, the memory 21._____
capacity would be rated
a) 1250 bits.   c) 3260 bits.
b) 20,000 bits.   d) 156 bits.

22. In the sign bit position, a 1 indicates a(n)   22._____
a) negative number.   c) octal code.
b) positive number.   d) hexadecimal code.

23. The 2s complement form of a binary number is the binary   23._____
number that results when:
a) all the 1s are changed to 0s.   c) 1 is added to the 1s complement.
b) all the 0s are changed to 1s.   d) both a and b

24. The ASCII code   24._____
a) is used with absolute encoders.
b) is considered to be an error-minimizing code.
c) includes letters as well as numbers.
d) all of the above

25. A(n) _____ bit is used to detect errors that may occur while 25._____
a word is moved.
a) parity   c) positive
b) negative   d) overflow

TEST 3.2

Place the answers to the following questions in the answer column at the right.

1. PLCs work on _____ numbers in one form or another to represent various codes or quantities.

1._____

2. The decimal system uses the number 9 as its base. (True or False)

2._____

3. The only allowable digits in the binary system are (a) _____ and (b) _____.

3a._____
3b._____

4. Each digit of a binary number is known as a(n) _____.

4._____

5. With reference to processor memory locations, the term *register* is often used interchangeably with ____ .

5._____

6. All digital computing devices perform operations in binary. (True and False)

6._____

7. The base of a number system determines the total number of unique symbols used by that system. (True or False)

7._____

8. Match the following bases with the appropriate number system.

BASE	NUMBER SYSTEM	
1) Base 2	a) Binary	8a._____
2) Base 16	b) Decimal	8b._____
3) Base 10	c) Octal	8c._____
4) Base 8	d) Hexadecimal	8d._____

9. In any number system, the position of a digit that represents part of the number has a weighted value associated with it. (True or False)

9._____

10. Match the following decimal numbers with their binary equivalent.

DECIMAL NUMBER	BINARY EQUIVALENT	
1) 9	a) 110011	10a._____
2) 37	b) 1001	10b._____
3) 51	c) 100101	10c._____
4) 42	d) 101010	10d._____

11. Usually a group of 8 bits is a byte, and a group of one or more bytes is a word. (True or False)

11._____

12. The _____bit of a word is the digit that represents the smallest value.

12._____

13. A memory that has a capacity of 700 sixteen-bit words can actually store _____ bits of information.

13._____

14. To express a number in binary requires fewer digits than in the decimal system. (True or False)

14._____

15. The octal number system consists of digits 0, 1, 2, 3, 4, 5, 6, and 7. There are no 8s or 9s. (True or False)

15._____

16. The octal number 46 expressed as a decimal number would be _____ .

16._____

17. The octal number 153 expressed as a binary number would be _____ .

17._____

18. The hexadecimal number system consists of 16 digits including the numbers 0 through 9 and letters A through F. (True or False)

18._____

19. Hexadecimal 2F equals _____ in decimal.

19._____

20. Hexadecimal A6 equals _____ in binary.

20._____

21. The decimal number 29 equals (a) _____ in binary and (b) _____in BCD.

21a._____

21b._____

22. The BCD number 1000 0101 0110 0111 equals _____ in decimal.　　22._____

23. In the Gray code there is a maximum of one bit change between　　23._____
two consecutive numbers. (True or False)

24. The radix of a number system is the same as the base. (True or False)　　24._____

25. Binary number systems use positive and negative symbols to　　25._____
represent the polarity of a number. (True or False)

26. Two systems of parity are normally used: (a) _____ and (b) ____ .　　26a._____
　　26b._____

27. Add binary 11101 and 1100.　　27._____

28. Subtract binary 11101 from 111010.　　28._____

29. Multiply binary 110 and 111.　　29._____

30. Divide binary 11010 by 10.　　30._____

31. The three basic compare instructions are (a) ___, (b) ___, and (c) ___.　　31a._____
　　31b._____
　　31c._____

Programming Assignments

1) Complete the following table using the change radix function of a PLC or online conversion calculator.

Binary	Octal	Decimal	Hexadecimal
101			
	11		
		15	
			D
	16		
1001011			
	47		
		73	

CHAPTER 4   Fundamentals of Logic

TEST 4.1

Choose the letter that best completes the statement.

1. The binary concept makes use of the fact that certain information   1._____
a) can exist in one of two possible states.
b) can be broken down into smaller units for easier analysis.
c) can be divided into two or more categories.
d) can be divided into two, or multiples of two, categories.

2. A gate is a device that   2._____
a) allows current flow in one direction only.
b) changes alternating current to direct current.
c) performs a logical decision based on its inputs.
d) performs a logical decision based on its outputs.

3. In conventional logic circuits, binary 1 represents   3._____
a) the presence of a signal.   c) a high voltage level.
b) the occurrence of some event.   d) all of these.

4. The logic function(s) used by PLCs is (are)   4._____
a) AND.   c) NOT.
b) OR.   d) all of these

5. The basic rule for an AND gate is   5._____
a) if all inputs are 1, the output will be 1.
b) if all inputs are 1, the output will be 0.
c) if all inputs are 0, the output will be 1.
d) both a and b

6. The basic rule for an OR gate is

6._____

a) if one or more inputs are 1, the output is 1.

b) if one or more inputs are 1, output is 0.

c) if one or more inputs are 0, the output is 1.

d) both b and c

7. The NOT function can be thought of as

7._____

a) a FALSE-to-TRUE converter.   c) an inverter.

b) a changer of states.   d) all of these.

8. A NOT function is used when a logic 1 must _____some device.

8._____

a) activate   c) switch

b) deactivate   d) light

9. The OR function, implemented using contacts, requires contacts

9._____

connected in

a) series.   c) series/parallel.

b) parallel.   d) parallel/series.

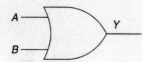
Figure 4-1 Logic symbol for question 10.

10-1. The logic symbol drawn in Figure 4-1 is that of the

10-1.____

a) AND function.   c) NOT function.

b) OR function.   d) NAND function.

10-2. The Boolean equation for the logic symbol is

10-2.____

a) $Y = A + B$ b) $Y = AB$

c) $Y = A \cdot B$ d) either b or c.

 **Figure 4-2 Logic symbol for question 11.**

11-1. The logic symbol drawn in Figure 4-2 is that of the   11-1._____

a) AND function.   c) NOT function.

b) OR function.   d) NOR function.

11-2. The Boolean equation for the logic symbol is   11-2._____

a) $Y = A + B + C$ c) $Y = (AB) + C$

b) $Y = ABC$ d) $Y = (A – B)C$

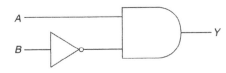

Figure 4-3 Logic symbol for question 12.

12. With reference to the logic circuit of Figure 4-3, the output Y will be   12._____

at a logic 1 when

a) inputs A and B are logic 1.

b) input A or B is logic 1.

c) input A is at logic 1 and input B is at logic 0.

d) input A is at logic 0 and input B is at logic 1.

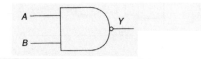

Inputs	Output
A B	Y
0 0	1
0 1	0
1 0	0
1 1	0
(a)	

Inputs	Output
A B	Y
0 0	0
0 1	1
1 0	1
1 1	1
(c)	

Figure 4-4 Logic symbol for question 13.

Inputs	Output
A B	Y
0 0	1
0 1	1
1 0	1
1 1	0
(b)	

Inputs	Output
A B	Y
0 0	0
0 1	0
1 0	0
1 1	1
(d)	

13-1. The logic symbol drawn in Figure 4-4 is that of the   13-1._____

a) AND function.   c) NOR function.

b) OR function.   d) NAND function.

13-2. The truth table for the logic symbol is_____.   13-2.____

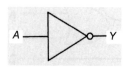

Figure 4-5 Logic symbol for question 14.

14-1. The logic symbol drawn in Figure 4-5 is that of the   14-1.____
a) NOT function.   c) NAND function.
b) NOR function.   d) OR function.

14-2. The Boolean equation for the logic symbol is: 14-2.____

a) $A = Y$ c) $\overline{A} = \overline{Y}$

b) $A = B$ d) $A = \overline{Y}$

Figure 4-6 Logic circuit for question 15.

15. The Boolean expression for the logic circuit drawn in Figure 4-6 is   15.____
a) $Y = ABC$ c) $Y = (A + B)C$
b) $Y = A + B + C$ d) $Y = AB + C$

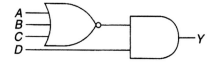

Figure 4-7 Logic circuit for question 16.

16. The Boolean expression for the logic circuit drawn in Figure 4-7 is   16.____
a) $Y = (AB)(CD)$ b) $Y = (\overline{A + B + C})D$

c) $Y = \overline{ABC} + D$ d) $Y = ABC + D$

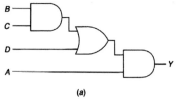
(a)

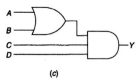
(c)

Figure 4-8 Logic circuits for question 17.

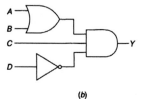
(b)

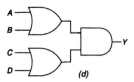
(d)

46

17. Which logic circuit of Figure 4-8 represents the Boolean expression   17._____
$Y = A(BC + D)$?

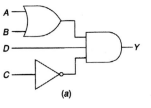
(a)

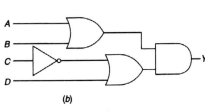
(b)

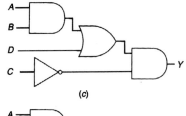
(c)

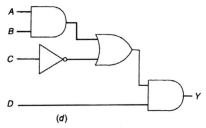
(d)

Figure 4-9 Logic circuits for question 18.

18. Which logic circuit of Figure 4-9 represents the Boolean expression   18._____
$Y=(A+B)(\overline{C}+D)$?

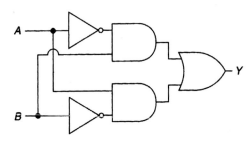

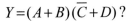

Figure 4-10 Logic circuit for question 19.

19. The Boolean expression for the logic circuit drawn in Figure 4-10 is   19._____

a) $Y=\overline{AB}+AB$

b) $Y=(\overline{AB})(A\overline{B})$

c) $Y=\overline{A}B+A\overline{B}$

d) $Y=(\overline{A}+B)(A+\overline{B})$

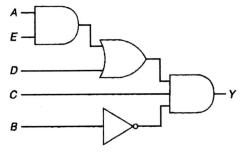

Figure 4-11 Logic circuit for question 20.

20. The Boolean expression for the logic circuit drawn in Figure 4-11 is   20._____

a) $Y=(A+E)DC\overline{B}$

b) $Y=AE(\overline{D}+C+\overline{B})$

c) $Y=(AE+D)C\overline{B}$

d) $Y=(A+E)\,\overline{DCB}$

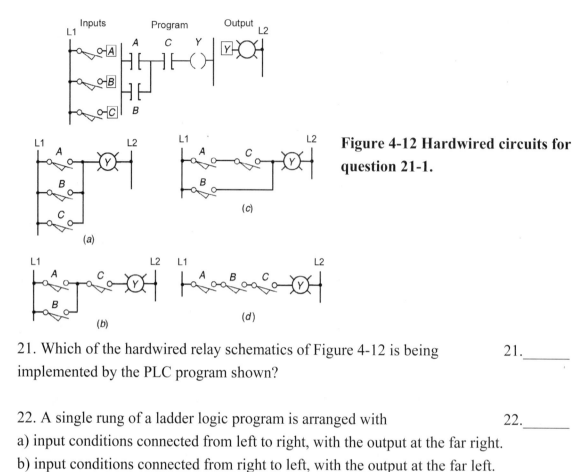

Figure 4-12 Hardwired circuits for question 21-1.

21. Which of the hardwired relay schematics of Figure 4-12 is being implemented by the PLC program shown?

21._____

22. A single rung of a ladder logic program is arranged with

22._____

a) input conditions connected from left to right, with the output at the far right.

b) input conditions connected from right to left, with the output at the far left.

c) the output in the center and the input conditions to the left and right of it.

d) all input conditions in parallel and all output conditions in series.

23. An AND gate operates on the same principle as

23._____

a) a series circuit.   c) a series-parallel circuit.

b) a parallel circuit.   d) none of these.

24. An OR gate operates on the same principle as

24._____

a) a series circuit.   c) a series-parallel circuit.

b) a parallel circuit.   d) none of these

25. A NOR gate is

25._____

a) an AND gate with an inverter connected to the output.

b) an OR gate with an inverter connected to the output.

c) equivalent to a series circuit.

d) equivalent to a parallel circuit.

26. The basic rule for an XOR function is

26._____

a) if one or the other, but not both, inputs are 1, the output is 1.

b) if one or more inputs are 1, the output is 1.

c) if one or more inputs are 1, the output is 0.

d) if one or more inputs are 0, the output is 1.

27. If you want to know when one or both matching bits in two different words are ON, you would use the _____ logic instruction.

27._____

a) AND

c) OR

b) XOR

d) NOT

TEST 4.2

Place the answers to the following questions in the answer column at the right.

1. The binary concept used in logic refers to the fact that many things can be thought of as existing in one of _____ states.

1._____

2. Normally, a binary 1 represents the presence of a signal, while a binary 0 represents the absence of a signal. (True or False)

2._____

3. A light that is ON or a switch that is closed would normally be represented by a binary _____ .

3._____

4. All gates are devices that have one input with which they perform logic decisions and produce a result at one or more of their outputs. (True or False)

4._____

5. The _____ gate output is 1 only if all inputs are 1.

5._____

6. The _____ gate output is 1 if one or more of its inputs are 1.

6._____

7. The NOT output is 1 if the input is _____.

7._____

8. The NOT function is also called a(n) _____.

8._____

9. In a NAND gate, when all inputs are 0, the output is _____.

9._____

10. In a two-input OR gate, when one input is 0 and the other one is 1, the output is _____.

10._____

11. In a NOR gate, when all inputs are 0, the output is _____.

11._____

12. In a two-input XOR gate, when both inputs are 0, the output is _____.

12._____

13. In a two-input XOR gate, when one input is 0 and the other is 1, the output is _____.

13._____

50

14. All inputs to an AND gate must be 1 to produce a 1 output.   14._____
(True or False)

15. All inputs to a NAND gate must be 1 to produce a 1 output.   15._____
(True or False)

16. Only one input to an OR gate must be 1 to produce a 1 output.   16._____
(True or False)

17. All inputs to a NOR gate must be 1 to produce a 1 output.   17._____
(True or False)

18. Inverting the output of an OR gate will result in creating a NOR gate.   18._____
(True or False)

19. The mathematical study of the binary number system and logic   19._____
is called _____ algebra.

20. The AND function, implemented using switches, will mean switches   20._____
connected in parallel. (True or False)

21. A two-input OR function, expressed as a Boolean equation, would be   21._____
$Y = AB$. (True or False)

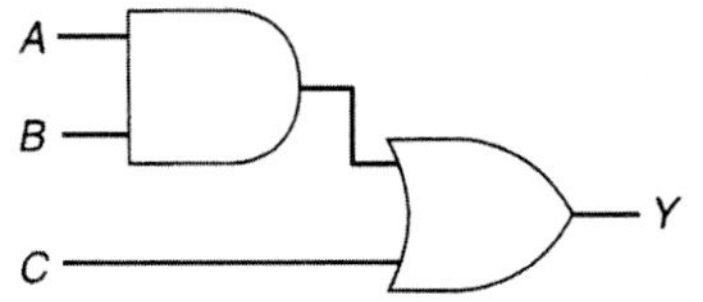

Figure 4-13 Logic circuit for question 22.

22. The correct Boolean expression for the logic circuit of   22._____
Figure 4-13 is _____.

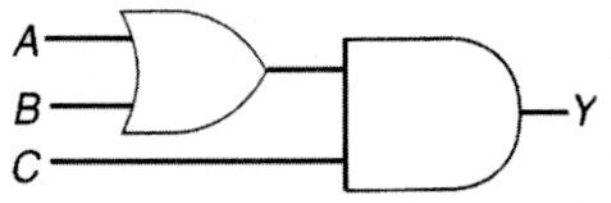

Figure 4-14 Logic circuit for question 23.

23. The correct Boolean expression for the logic circuit of   23._____
Figure 4-14 is _____.

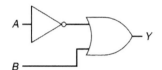 **Figure 4-15 Logic circuit for question 24.**

24. The correct Boolean expression for the logic circuit of   24._____
Figure 4-15 is ____.

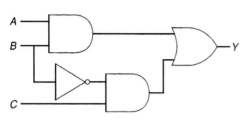 **Figure 4-16 Logic circuit for question 25.**

25. The correct Boolean expression for the logic circuit of   25._____
Figure 4-16 is ____.

26. Hardwired logic refers to logic control functions determined by the   26._____
way devices are interconnected. (True or False)

27. Hardwired logic can be implemented using relays and relay   27._____
schematics. (True or False)

28. Hardwired logic is fixed and is changeable only by altering the   28._____
way devices are connected. (True or False)

29. Programmable control is based on logic functions that are   29._____
programmable and easily changed. (True or False)

30. There is no difference between a relay schematic and a ladder logic   30._____
program. (True or False)

31. On some PLCs, only one output is allowed per ladder logic rung.   31._____
(True or False)

32. One of the most common PLC programming languages is ladder logic.   32._____
(True or False)

33. Ladder logic is a graphical representation of a user program.   33._____
(True or False)

34. Complete the truth table of basic Boolean operations by signifying the correct true or false condition for each blank space.

A	B	A and B	A or B	not A	A xor B
False	False				
False	True				
True	False				
True	True				

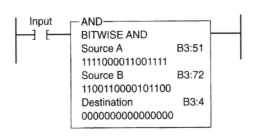

Figure 4-17 Instruction for question 35.

35. What will be the data stored in the destination address B3:4 of   35._____
Figure 4-17 when the input is true?

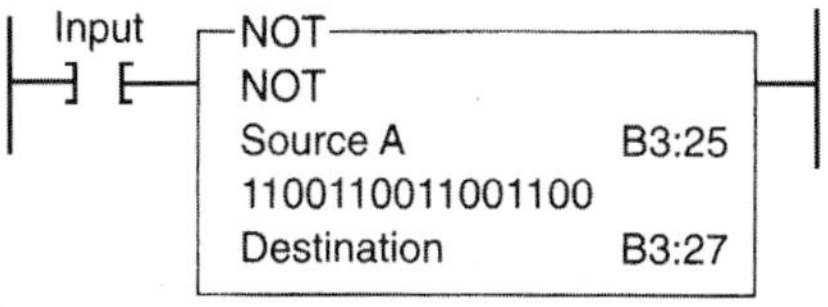

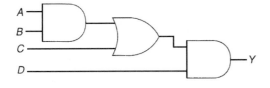

Figure 4-18 Instruction for question 36.

36. What will be the data stored in the destination address B3:27 of   36._____
Figure 4-18 when the input is true?

Figure 4-19 Gate logic for question 37 program answer.

37. Draw a PLC ladder diagram program for the gate logic array shown in Figure 4-19.

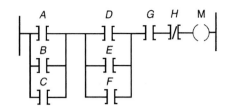

Figure 4-20 PLC ladder diagram logic gate array answer for question 38.

38. Draw the equivalent gate logic array for the PLC ladder diagram shown in Figure 4-20.

Programming Assignments

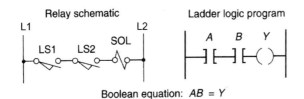

Figure 4-21 Program for assignment 1.

Boolean equation: $AB = Y$

1) Program the relay schematic of Figure 4-21 using your PLC, and simulate its operation.

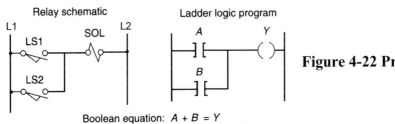

Figure 4-22 Program for assignment 2.

Boolean equation: $A + B = Y$

2) Program the relay schematic of Figure 4-22 using your PLC, and simulate its operation.

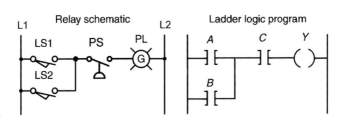

Figure 4-23 Program for assignment 3.

Boolean equation: $(A + B)C = Y$

3) Program the relay schematic of Figure 4-23 using your PLC, and simulate its operation.

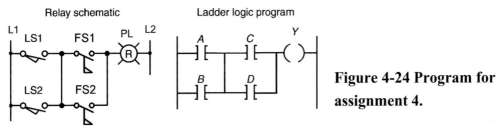

Figure 4-24 Program for assignment 4.

Boolean equation: $(A + B)(C + D) = Y$

4) Program the relay schematic of Figure 4-24 using your PLC, and simulate its operation.

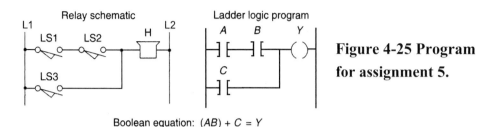

Figure 4-25 Program for assignment 5.

Boolean equation: $(AB) + C = Y$

5) Program the relay schematic of Figure 4-25 using your PLC, and simulate its operation.

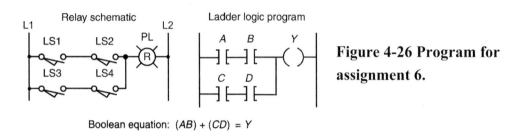

Figure 4-26 Program for assignment 6.

Boolean equation: $(AB) + (CD) = Y$

6) Program the relay schematic of Figure 4-26 using your PLC, and simulate its operation.

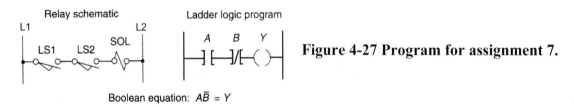

Figure 4-27 Program for assignment 7.

Boolean equation: $A\overline{B} = Y$

7) Program the relay schematic of Figure 4-27 using your PLC, and simulate its operation. The program is meant to turn on the output when LS1 is closed and LS2 is not closed.

8) Execute each of the following Boolean equations as a ladder logic rung. Program each rung into the PLC, and prove its operation.

a) $Y = (A+B)CD$

c) $Y = [(\overline{A}+\overline{B})C] + D\overline{E}$

b) $Y = (A\overline{B}C) + \overline{D} + E$

d) $Y = (AB\overline{C}) + (D\overline{E}F)$

9) Develop a PLC program that will simulate the operation of the XOR function. Enter the program into the PLC, and prove its operation.

10) A conveyor will run when any one of four inputs is on. It will stop when any one of four other inputs is on. Develop a PLC program that will simulate this operation. Enter the program into the PLC, and prove its operation.

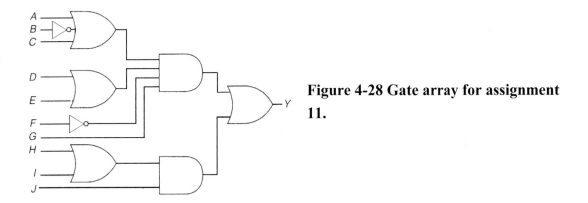

Figure 4-28 Gate array for assignment 11.

11) Develop a PLC program that will simulate the gate array logic shown in Figure 4-28. Enter the program into the PLC, and prove its operation.

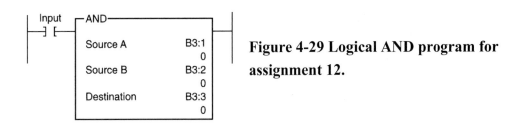

Figure 4-29 Logical AND program for assignment 12.

12) Enter the logical AND program shown in Figure 4-29 into the PLC. Use the data monitor function to store the following data:

B3:1 = 1111 0000 1111 0000   B3:2 = 0000 0000 1111 0000

Run the program, and verify that B3:3 contains the following bit pattern:

B3:3 = 0000 0000 1111 0000

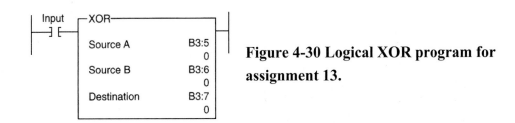

Figure 4-30 Logical XOR program for assignment 13.

13) Enter the logical XOR program shown in Figure 4-30 into the PLC. Use the data monitor function to store the following data:

B3:5 = 1010 1100 0111 1111   B3:6 = 1010 0111 0111 0111

Run the program, and verify that B3:7 contains the following bit pattern:

B3:7 = 0000 1011 0000 1000

14) Assume you have an alarm condition as part of a machine operation that tells you when one (or more) of eight limit switches wired into one input module is not in the correct position to allow the process to continue. Develop a program that uses the XOR logical instruction and that provides a simple means for the troubleshooter to isolate which switch (or switches) is in the wrong position by operating a pushbutton. Enter the program into the PLC, and prove its operation.

CHAPTER **5**   **Basics of PLC Programming**

TEST 5.1

Choose the letter that best completes the statement.

1. The _____ will account for most of the total memory of a given   1.___C___
PLC system.
a) input image table file   c) user program
b) output image table file   d) internal operating instructions

2. The status bit of switches and pushbuttons connected to a PLC are   2.___A___
stored in the
a) input image table file.   c) user program.
b) output image table file.   d) all of these.

3. The memory organization of a PLC can be divided into what two broad   3.___C___
categories?
a) Input and output image files   c) Program and data files
b) Timer and counter files   d) Control and integer files

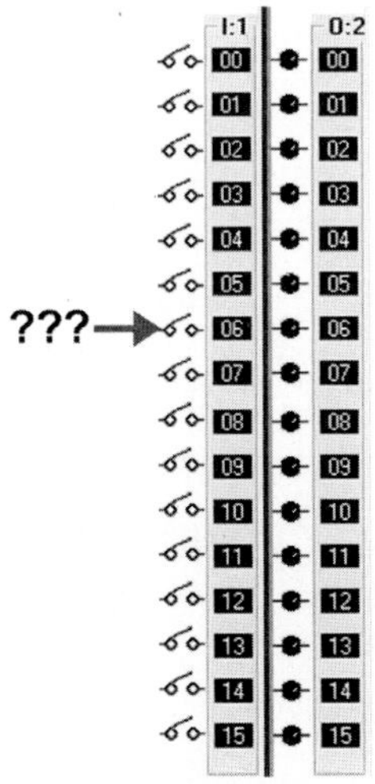

Figure 5-1 I/O module for question 4.

4. The address for the point on the I/O module shown in Figure 5-1   4.___B___
would be
a) I:6/1.   c) O:6/1.
b) I:1/6.   d) O:1/6.

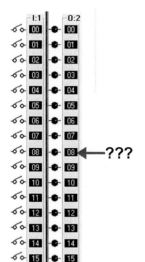

Figure 5-2 I/O module for question 5.

???

5. The address for the point on the I/O module shown in Figure 5-2 would be

5. <u>D</u>

a) I:2/8.

c) O:8/2.

b) I:8/2.

d) O:2/8.

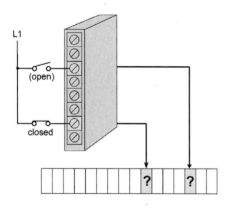

Figure 5-3 Input module for question 6.

6. For the input module of Figure 5-3, the data stored in the word corresponding to the open switch would be ___ and that for the closed switch would be ___, respectively.

6. <u>B</u>

a) 1, 1

c) 0, 0

b) 0, 1

d) 1, 0

60

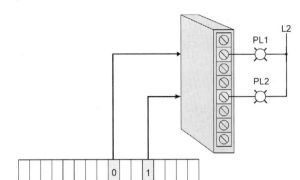

Figure 5-4 Output module for question 7.

7. For Figure 5-4, according to the data stored in the word corresponding to the outputs, PL1 would be _____ and PL2 would be ___, respectively.

7. _D_

a) on, on

c) on, off

b) off, off

d) off, on

8. The scan is normally a sequential process of

8. _D_

a) reading the control logic, evaluating the outputs, and updating the inputs.

b) writing the control logic, evaluating the outputs, and updating the inputs.

c) reading/writing the status of inputs and updating the outputs.

d) reading the status of inputs, evaluating the control logic, and energizing or de-energizing the outputs.

9. Which of the following is a factor in determining the total scan time?

9. _D_

a) Length of the ladder program

b) Type of instructions executed

c) Speed of the processor

d) All of these

10. If a PLC has a total scan time of 10 ms and has to monitor a signal that _____, then the controller may not detect this change.

10. _D_

a) changes state once in 20 ms

c) is constantly changing

b) is fast

d) changes state twice in 5 ms

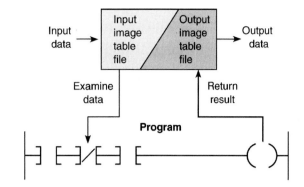

Figure 5-5 Scan process for question 11.

11-1. For the scan process illustrated in Figure 5-5, the input data are provided by the

11-1. _C_

a) ladder program.

c) input module.

b) output module.

d) all of these.

11-2. The output data are sent to the

11-2. _A_

a) output field devices.

c) input module.

b) output module.

d) ladder program.

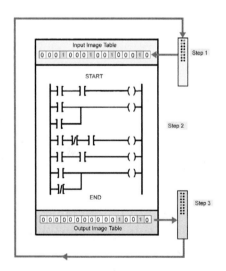

Figure 5-6 Scan process for question 12.

12. For the scan process illustrated in Figure 5-6, step 2 involves

12. _A_

a) solving the ladder program.

c) transferring data to the input module.

b) transferring data to the output module.

d) reading data from the input module.

13. The two types of patterns used to accomplish the scan function are

13. _A_

a) horizontal and vertical.

c) up and down.

b) left and right.

d) input and output.

14. A(n) _____ scan pattern examines instructions rung by rung.   14. _A_

a) horizontal   c) input
b) vertical   d) output

15. The actual scan time is: 15. _D_

a) calculated and stored in the PLC's memory.
b) computed each time the END instruction is executed.
c) the time taken to scan inputs and outputs and execute the user program.
d) all of these

16. Which of the following standard PLC programming languages is a   16. _D_
graphical language?

a) Function Block Diagram   c) Sequential Function Chart
b) Ladder Diagram   d) All of these

BAND_01

BAND
Boolean And

Caution

0

Out — PL 1

Sensor 1 — 0 — In1
Sensor 2 — 0 — In2

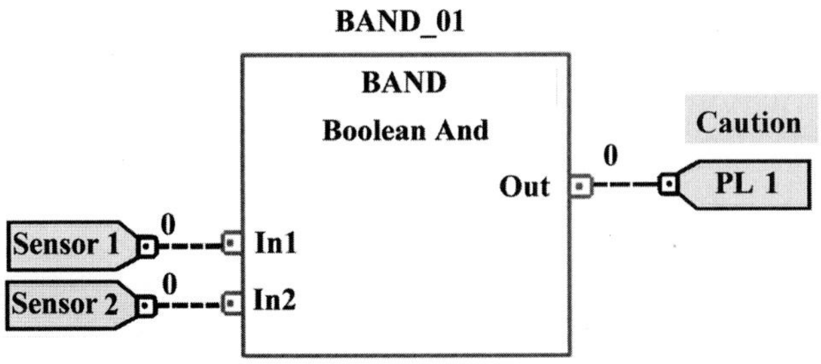

Figure 5-7 Program for question 17.

17. The PLC program of Figure 5-7 is a ___ type.   17. _B_

a) structured text   c) ladder diagram
b) functional block diagram   d) structured text

18. The Examine If Closed instruction   18. _B_

a) is also known as the Examine-On instruction.
b) is also known as the XIC instruction.
c) looks and operates like a normally open relay contact.
d) all of these

19. The ____ instructions always interpret a 1 status as true and a 0 status   19. _A_
as false.

a) XIC   c) contact
b) XIO   d) all of these

20. The Examine If Open instruction

a) is also known as the Examine-Off instruction.

b) is also known as the XIO instruction.

c) looks and operates like a normally closed relay contact.

d) all of these

20. _D_

21. The Output Energize instruction

a) is also known as the OTE instruction.

b) signals the PLC to energize or deenergize the output.

c) looks and operates like a relay coil.

d) all of these

21. _D_

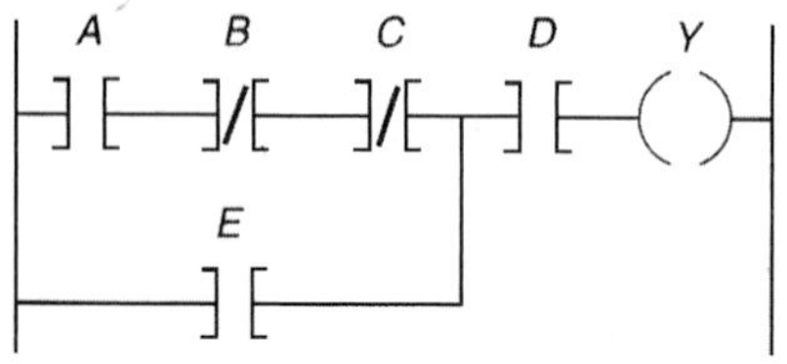

Figure 5-8 Program for question 22.

22-1. For the program of Figure 5-8, instruction *A* represents an

a) Examine If Closed instruction.　　c) Output Energize instruction.

b) Examine If Open instruction.　　d) Input Energize instruction.

22-1. _A_

22-2. Instruction *B* represents an

a) Examine If Closed instruction.　　c) Output Energize instruction.

b) Examine If Open instruction.　　d) Input Energize instruction.

22-2. _B_

22-3. Instruction *Y* represents an

a) Examine If Closed instruction.　　c) Output Energize instruction.

b) Examine If Open instruction.　　d) Input Energize instruction.

22-3. _C_

22-4. Which of the following input combinations will result in the *Y* output being energized?

a) A and *B* and C and *D*　　c) *A* or *B* or *C* or *D*

b) A and not *B* and not C and *D*　　d) *E* and not *D*

22-4. _B_

23. A programmed XIO instruction   23. ___D___

a) with a bit status of 1 will not have logic continuity.

b) is examined for an OFF condition.

c) is examined for an ON condition.

d) both a and b

24. A programmed XIC instruction   24. ___A___

a) with a bit status of 1 will have logic continuity.

b) with a bit status of 0 will have logic continuity.

c) is examined for an OFF condition.

d) both b and c

25. A ladder rung is said to have logic continuity when   25. ___A___

a) at least one left-to-right true logical path exists.

b) all input instructions are at a logic 1.

c) all input instructions are at a logic 0.

b) both a and b

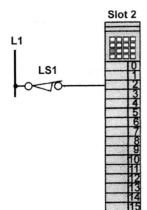

Figure 5-9 Module for question 26.

26. The most likely module address for LS1 of Figure 5-9 is   26. ___C___

a) LS1 I2.   c) I:2/2.

b) LS1 O2.   d) O:2/2.

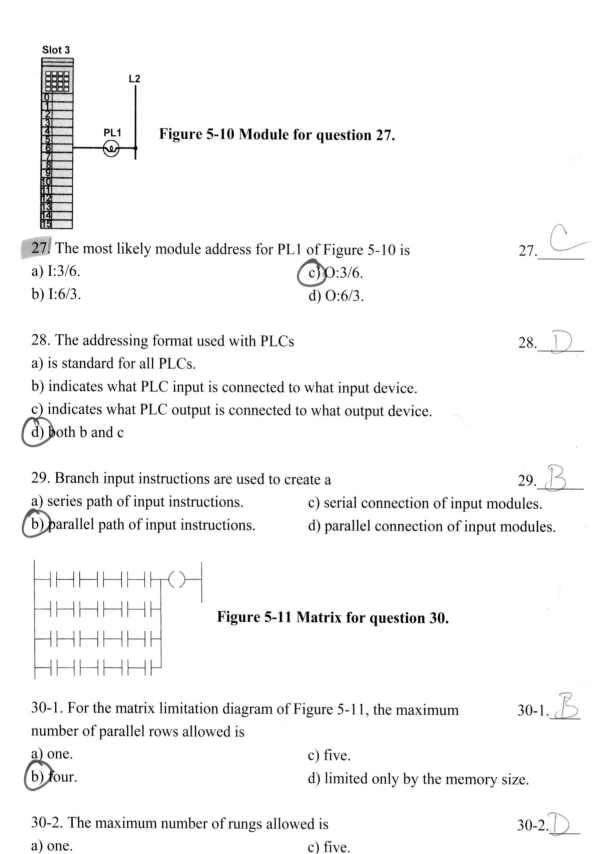

Slot 3

L2

PL1

Figure 5-10 Module for question 27.

27. The most likely module address for PL1 of Figure 5-10 is   27. _C_
a) I:3/6.   c) O:3/6.
b) I:6/3.   d) O:6/3.

28. The addressing format used with PLCs   28. _D_
a) is standard for all PLCs.
b) indicates what PLC input is connected to what input device.
c) indicates what PLC output is connected to what output device.
d) both b and c

29. Branch input instructions are used to create a   29. _B_
a) series path of input instructions.   c) serial connection of input modules.
b) parallel path of input instructions.   d) parallel connection of input modules.

Figure 5-11 Matrix for question 30.

30-1. For the matrix limitation diagram of Figure 5-11, the maximum   30-1. _B_
number of parallel rows allowed is
a) one.   c) five.
b) four.   d) limited only by the memory size.

30-2. The maximum number of rungs allowed is   30-2. _D_
a) one.   c) five.
b) four.   d) limited only by the memory size.

66

30-3. The maximum number of series contacts allowed per rung is   30-3. _C_

a) one.   c) five.

b) four.   d) limited only by the memory size.

30-4. The maximum number of outputs allowed per rung is   30-4. _A_

a) one.   c) five.

b) four.   d) limited only by the memory size.

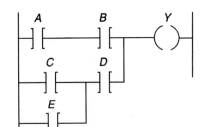

Figure 5-12 Program for question 31.

31. In Figure 5-12, the nested contact is contact   31. _D_

a) B.   c) D.

b) C.   d) E.

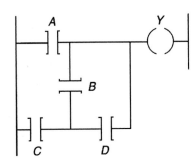

Figure 5-13 Program for question 32-1.

32-1. The Boolean equation for the logic represented in the ladder logic   32-1. _A_
program in Figure 5-13 can be expressed as

a) Y = (A) + (CD) + (BC)   c) Y = (AB) + (CD)

b) Y = A + B + (CD)   d) Y = A(BCD)

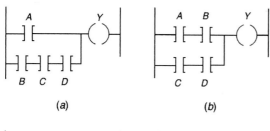

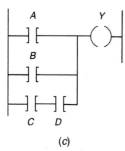

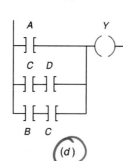

(a) (b)

(c) (d)

Figure 5-14 Programs for question 32-2.

32-2. The ladder logic program can be reprogrammed as shown in
Figure 5-14 to eliminate the vertical programmed contact and maintain
the same input logic conditions.

32-2. _D_

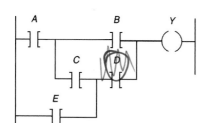

Figure 5-15 Program for question 33-1.

33-1. With reference to the program of Figure 5-15, if it could be
programmed as shown, part of the logic would be ignored due to the fact
that the processor allows for a flow

33-1. _B_

a) from right to left only.   c) in the upward direction only.
b) from left to right only.   d) both a and c

33-2. The Boolean equation for the logic represented in the ladder diagram
can be expressed as

33-2. _B_

a) $Y = (AB) + (ACD) + (DE)$.   c) $Y = (AB) + (AC) + (AD) + (ED)$.
b) $Y = (AB) + (ACD) + (DE) + (BCE)$.   d) $Y = (AB) + (CD) + E$.

68

34. An internal control relay   34. __A__

a) does not directly control an output field device.

b) is not controlled by the programmed logic.

c) is used primarily to control the internal power to the processor module.

d) is used primarily when controlling multiple output circuits.

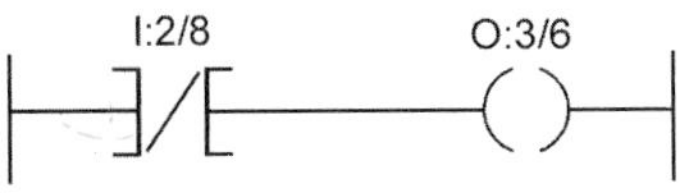

Figure 5-16 Ladder rung for question 34.

35. In Figure 5-16, the bit status condition of the input device connected to   33. __A__
address I:2/8 must be _____ to turn on output address O:3/6.

a) 0   c) normally open

b) 1   d) normally closed

36. The address I:1/3 identifies an input module residing in slot ___   36. __B__
of the PLC chassis.

a) zero   c) two

b) one   d) three

37. The address O:3/7 would be found on terminal _____   37. __B__
of an output module residing in slot ___ of the PLC chassis.

a) 3, 7   c) 4, 7

b) 7, 3   d) 7, 4

38. When using an XIC instruction, the reference address could be   38. __D__

a) a bit from an input device.   c) a bit from an internal relay.

b) a bit from an output device.   d) any of these

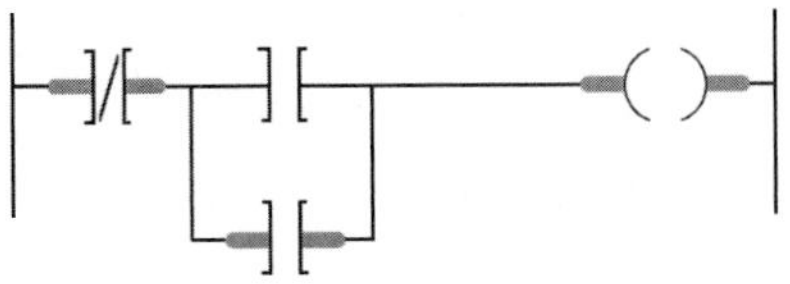

Figure 5-17 Ladder rung for question 39.

39. The highlighted rungs in the program rung shown in Figure 5-17   39. __A__
indicate

a) the instruction is true.

b) the instruction is false.

c) the instruction does not have logic continuity.

d) both b and c

69

40. When the PLC is required to operate the user program without energizing 40. _D_
any outputs, it is placed in the ___ mode.
a) RUN   c) PROGRAM
b) CLEAR MEMORY   d) TEST

41. In ladder logic programs, outputs are represented by   41. _B_
a) contact symbols.   c) schematic load device symbols.
b) coil symbols.   d) either a or b.

42. When a program rung consists of an output instruction only, the output   42. _A_
would be
a) continuously ON.   c) shorted.
b) continuously OFF.   d) both a and c.

43. Parallel connections of ladder logic are typically called   43. _D_
a) rungs.   c) coils.
b) networks.   d) branches.

44. Each complete horizontal line of a ladder diagram is generally referred   44. _A_
to as a(n)
a) rung.   c) input.
b) branch.   d) output.

45. The last element to be entered on a ladder rung is a(n)   45. _A_
a) coil.   c) XIC instruction.
b) contact.   d) XIO instruction.

46. A normally open limit switch is wired to an input module and   46. _D_
programmed using an XIO instruction. The instruction will be true when
a) power is applied and the PLC is in the run mode.
b) the limit switch is closed.
c) the limit switch is open.
d) never

70

TEST 5.2

Place the answers to the following questions in the answer column at the right.

1. All PLC manufacturers organize their memories in the same way. (True or False)

1._____

2. The memory organization of a PLC is often called a memory map. (True or False)

2._____

3. Allen-Bradley PLCs have two different memory structures identified by the terms (a) ____-based and (b) ____-based systems.

3a._____
3b._____

4. The memory space can be divided into the two broad categories of (a) ____ files and (b) ____ files.

4a._____
4b._____

5. Match the following data files with the closest description. Place the name from the data file list in the answer column.

DATA FILE
1) Bit
2) Integer
3) Input
4) Status
5) Timer
6) Output

DESCRIPTION
a) Used for internal relay logic storage.
b) Used for storage of the status of input field devices.
c) Used for the storage of accumulated and preset values.
d) Stores controller operation information.
e) Used to store numeric values.
f) Used for storage of the status of output field devices.

5a._____
5b._____
5c._____
5d._____
5e._____
5f._____

6. Most of the total PLC memory is used for the ____.

6._____

7. The user program contains the logic that controls the machine operation. (True or False)

7._____

8. Most instructions require one word of memory. (True or False)

8._____

9. The address **I:1/4** breaks down into the following parts: I is for (a) ____. The colon is used to separate the module type from the (b) ____. One is the (c) ____ number. The forward slash is used to separate the slot from the (d) ____. Four is the (e) ____ number.

9a._____
9b._____
9c._____
9d._____
9e._____

10. The status of input and output devices is stored in a 1 data table. (True or False)

10._____

11. If a switch connected to an input module is closed, a binary 1 is stored in the proper ___ location.

11._____

12. The ___ image table file is updated during the I/O scan to reflect the current status of digital inputs.

12._____

13. If the program calls for a specific output to be on, the corresponding bit in the output image table file is set to ___ .

13._____

14. The program ___ cycle is a continuous and sequential process of reading the status of inputs, evaluating the control logic, and updating the outputs.

14._____

15. The greater the scan time, the faster the PLC can react to changes in inputs. (True or False)

15._____

16. Scan time varies with program content and length. (True or False)

16._____

17. If any input signal changes state very quickly, it is possible that the controller may never be able to detect the change. (True or False)

17._____

18. Scan patterns are identified as being either (a) ____ or (b) ____.

18a._____
18b._____

19. Misunderstanding the way the PLC scans a program can cause programming bugs. (True or False)

19._____

20. It takes the processor exactly the same amount of time to examine different types of instructions. (True or False)

20._____

21. The ladder logic program language is basically a(n) ___ set of instructions used to create the controller program.

21._____

22. (a) ___ (b) ___ are the basic symbols of the ladder logic program instruction set.

22a._____
22b._____

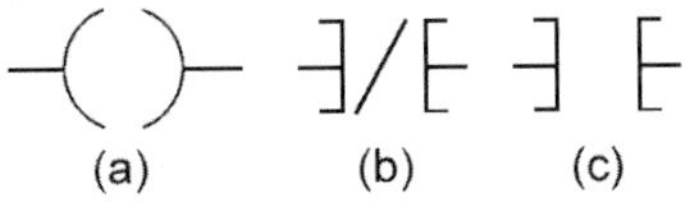

Figure 5-18 Instructions for question 23.

(a) (b) (c)

23. Identify the relay ladder logic instructions shown in Figure 5-18.

23a. _____
23b._____
23c._____

24. In general, a ladder logic rung consists of input conditions represented by (a) ____ symbol and an output instruction represented by the (b) ____ symbol.

24a._____
24b._____

25. When the XIC instruction is associated with a physical input, the instruction will be set to 1 when there is no input voltage applied to the terminal. (True or False)

25._____

26. When the XIO instruction is associated with a physical input, the instruction will be set to 1 when there is no input voltage applied to the terminal. (True or False)

26._____

27. Both normally open and normally closed pushbuttons can be examined for a XIC or XIO condition. (True or False)

27._____

28. The status of the OTE instruction is set to 1 to energize the output and to 0 to de-energize the output. (True or False)

28._____

29. When logic _____ exists in at least one path, the rung condition is said to be true.

29._____

30. The main function of the ladder logic program is to control outputs based on ___ conditions.

30._____

31. Each individual contact instruction can be used only once throughout the program. (True or False)

31._____

32. The rung condition and OTE instruction are false if no logical continuity path has been established. (True or False)

32._____

33. The addressing format for inputs and outputs is standard for all PLC models. (True or False)

33._____

34. The ___ will indicate what PLC input is connected to what input device and what PLC output will drive what output device.

34._____

35. There may be a limit to the number of series contact instructions that can be included in one rung of a ladder logic program. (True or False)

35._____

36. Branch instructions are used to create _____ paths of input conditions.

36._____

37. When there is a true logic rung path, all parallel outputs in the rung become true. (True or False)

37._____

38. A(n) ___branch starts or ends within another branch.

38._____

39. On some PLC models, branches can be established at both the input and output portions of a rung. (True or False)

39._____

40. An internal output does not directly control an output field device. (True or False)

40._____

41. An internal output is used when an output instruction is required but no connection to a(n) _____ is required.

41._____

42. Match the PLC modes of operation with the most correct description. Select the answers from the mode list, and place them in the answer column.

MODE
1) Program
2) Test
3) Run

DESCRIPTION

a) Used to execute the user program.   42a._____

b) Used to monitor the user program without energizing any outputs.   42b._____

c) Used to enter the user program.   42c._____

Programming Assignments

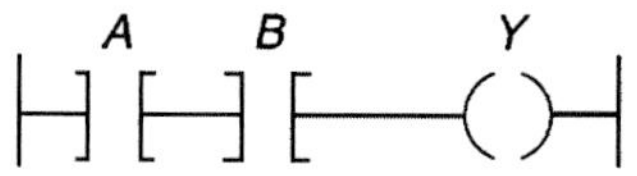

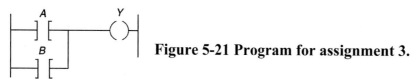

Figure 5-19 Program for assignment 1.

1a) On a separate sheet, redraw the ladder logic program of Figure 5-19 using addressing that applies to your PLC. Use normally open pushbuttons or switches for the input field devices and a pilot light for the output field device. Program the circuit into the controller, and verify its operation.

  b) What combination of the input conditions will result in an output at *Y*?

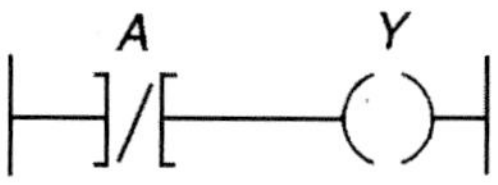

Figure 5-20 Program for assignment 2.

2a) On a separate sheet, redraw the ladder logic program of Figure 5-20 using addressing that applies to your PLC. Use a normally open pushbutton or switch for the input field device and a pilot light for the output field device. Program the circuit into the controller, and verify its operation.

  b) What input condition will result in an output at *Y*?

Figure 5-21 Program for assignment 3.

3a) On a separate sheet, redraw the ladder logic program of Figure 5-21 using addressing that applies to your PLC. Use normally open pushbuttons or switches for the input field devices and a pilot light for the output field device. Program the circuit into the controller, and verify its operation.

  b) What combination of the input conditions will result in an output at *Y*?

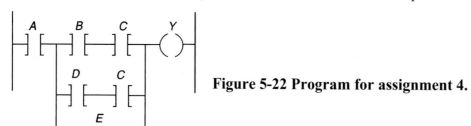

Figure 5-22 Program for assignment 4.

4a) On a separate sheet, redraw the ladder logic program of Figure 5-22 using addressing that applies to your PLC. Use normally open pushbuttons or switches for the input field devices and a pilot light for the output field device. Program the circuit into the controller, and verify its operation.

b) What combination of the input conditions will result in an output at *Y*?

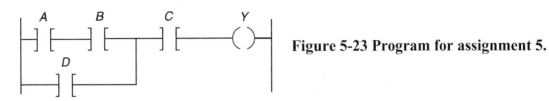

Figure 5-23 Program for assignment 5.

5a) On a separate sheet, redraw the ladder logic program of Figure 5-23 using addressing that applies to your PLC. Use normally open pushbuttons or switches for the input field devices and a pilot light for the output field device. Program the circuit into the controller, and verify its operation.

b) What combination of the input conditions will result in an output at *Y*?

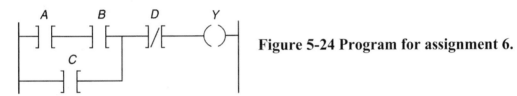

Figure 5-24 Program for assignment 6.

6a) On a separate sheet, redraw the ladder logic program of Figure 5-24 using addressing that applies to your PLC. Use normally open pushbuttons or switches for the input field devices and a pilot light for the output field device. Program the circuit into the controller, and verify its operation.

b) What combination of the input conditions will result in an output at *Y*?

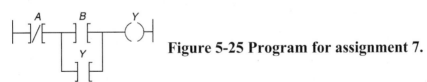

Figure 5-25 Program for assignment 7.

7a) On a separate sheet, redraw the ladder logic program of Figure 5-25 using addressing that applies to your PLC. Use normally open pushbuttons or switches for the input field devices and a pilot light for the output field device. Program the circuit into the controller, and verify its operation.

b) What combination of the input conditions will result in an output at *Y*?

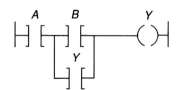

Figure 5-26 Program for assignment 8.

8a) On a separate sheet, redraw the ladder logic program of Figure 5-26 using addressing that applies to your PLC. Use normally open pushbuttons or switches for the input field devices and a pilot light for the output field device. Program the circuit into the controller, and verify its operation.

   b) What combination of the input conditions will result in an output at *Y*?

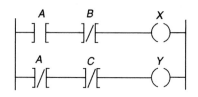

Figure 5-27 Program for assignment 9.

9a) On a separate sheet, redraw the ladder logic program of Figure 5-27 using addressing that applies to your PLC. Use normally open pushbuttons or switches for the input field devices and pilot lights for the output field devices. Program the circuit into the controller, and verify its operation.

   b) What combination of the input conditions will result in an output at *X* and *Y*?

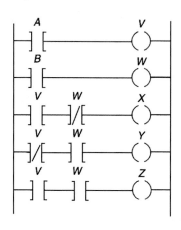

Figure 5-28 Program for assignment 10.

10a) On a separate sheet, redraw the ladder logic program of Figure 5-28 using addressing that applies to your PLC. Use normally open pushbuttons or switches for the input field devices and pilot lights for the output field devices. Program the circuit into the controller, and verify its operation.

   b) What combination of the input conditions will result in an output at *V, W, X, Y,* and *Z?*

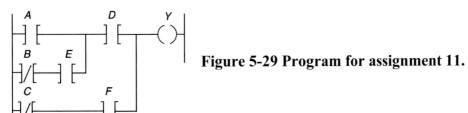

Figure 5-29 Program for assignment 11.

11a) On a separate sheet, redraw the ladder logic program of Figure 5-29 to maintain the original control logic and eliminate the nested branch within a branch. Use normally open pushbuttons or switches for the input field devices and a pilot light for the output field devices. Program the circuit into the controller, and verify its operation.

   b) What combination of the input conditions will result in an output at *Y*?

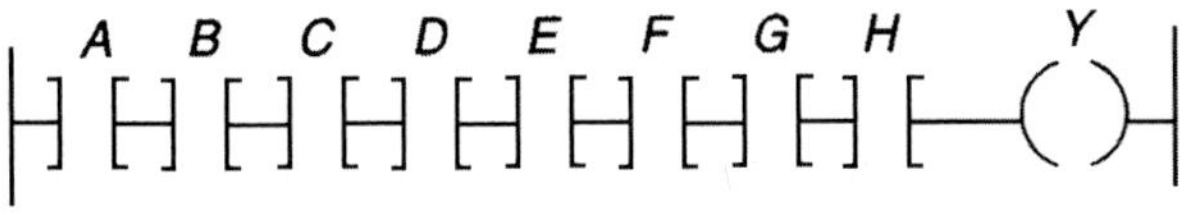

Figure 5-30 Program for assignment 12.

12) Assume that the PLC used to program the circuit in Figure 5-30 can accommodate a maximum of five series contact instructions per rung. On a separate sheet of paper, redesign the program to meet this PLC requirement by using an internal relay instruction. Program the circuit into the controller, and verify its operation.

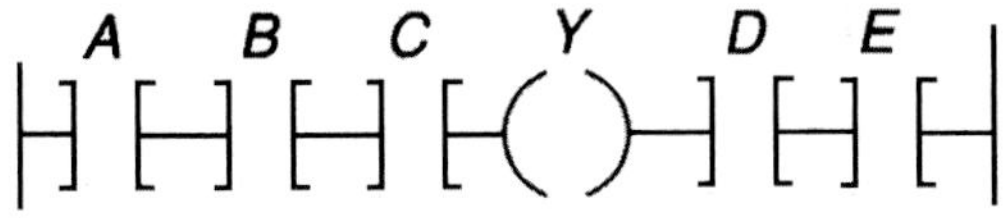

Figure 5-31 Program for assignment 13.

13a) What problem is posed in trying to program the ladder logic program of Figure 5-31 into a PLC?

   b) On a separate sheet, redraw the ladder logic program of Figure 5-31 to maintain the original control logic and eliminate the problem. Program the circuit into the controller, and verify its operation.

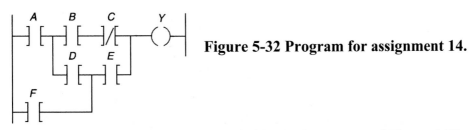

Figure 5-32 Program for assignment 14.

14a) On a separate sheet, redraw the ladder logic program of Figure 5-32 to solve the problem of some logic ignored. Program the circuit into the controller, and verify its operation.

   b) What combination of the input conditions will result in an output at *Y*?

CHAPTER 6

Developing Fundamental PLC Wiring Diagrams and Ladder Logic Programs

TEST 6.1

Choose the letter that best completes the statement.

1. An electromagnet control relay is basically a(n):　　　　　　　　　　1.＿＿＿＿＿

a) electromagnet used to switch contacts.

b) electromagnet used to relay information.

c) manually operated control device.

d) pressure-operated control device.

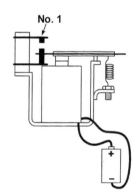

No. 1

Figure 6-1 Relay illustration for question 2.

2-1. In the relay illustration of Figure 6-1, the coil would be considered　　2-1.＿＿＿＿＿
to be:

a) energized.　　　　　　　　　　c) operated from an AC source.

b) deenergized.　　　　　　　　　d) both a and c.

2-2. In the relay illustration the contact No. 1 is a(n):　　　　　　　　2-2.＿＿＿＿＿

a) NO fixed contact.　　　　　　　　c) NO movable contact.

b) NC fixed contact.　　　　　　　　d) NC movable contact.

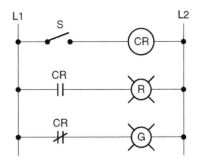

Figure 6-2 Hardwired relay circuit for question 3.

3. In the hardwired relay control circuit of Figure 6-2, when the switch is closed, CR coil is:

3._____

a) energized, and the red and green lights are both on.

b) deenergized, the red light is off, and the green light is on.

c) energized, the red light is on, and the green light is off.

d) energized, the red light is off, and the green light is on.

4. A contactor:

4._____

a) is another name for a relay.

b) is designed to handle heavy power loads.

c) always has an overload relay physically and electrically attached.

d) is a physically small relay.

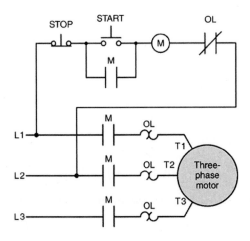

Figure 6-3 Motor starter circuit for question 5.

5-1. In the motor starter circuit of Figure 6-3, the main contacts M are: 5-1._____

a) part of the power circuit.

b) part of the control circuit.

c) designed to handle the full load current of the motor.

d) both a and c.

5-2. The motor starter coil M is: 5-2._____

a) part of the power circuit.

b) energized to start the motor.

c) energized only as long as the start button is pressed.

d) all of the above.

5-3. Any overload current is sensed by the: 5-3._____

a) starter coil.   c) OL coils.

b) control contact M.   d) OL contact.

6. The abbreviations NO (normally open) and NC (normally closed) 6._____

represent the electrical state of switch contacts when:

a) power is applied.   c) the switch is actuated.

b) power is not applied.   d) the switch is not actuated.

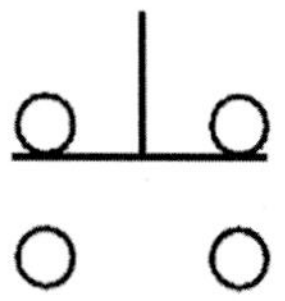

Figure 6-4 Pushbutton symbol for question 7.

7. The pushbutton symbol shown in Figure 6-4 would be classified as a(n): 7._____

a) NO pushbutton.   c) break-before-make pushbutton.

b) NC pushbutton.   d) ON/OFF pushbutton.

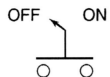

Figure 6-5 Symbol for question 8.

8. The device represented by the symbol of Figure 6-5 is a: 8._____

a) drum switch.   c) sequence switch.

b) selector switch.   d) toggle switch.

9. A limit switch is usually actuated by: 9._____

a) hand.   c) contact with an object.

b) pressure.   d) an electromagnet.

10. A proximity switch can be actuated: 10._____

a) without any physical contact.   c) by a change in capacitance.

b) by a change in light intensity.   d) by all of the these.

Figure 6-6 Symbol for question 11.

11. Figure 6-6 represents the symbol for a: 11._____

a) pressure switch.   c) limit switch.

b) temperature switch.   d) level switch.

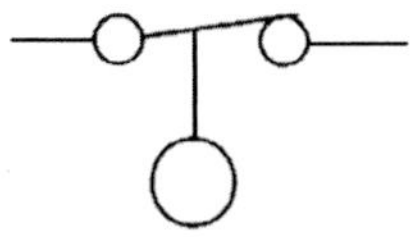

Figure 6-7 Symbol for question 12.

12. Figure 6-7 represents the symbol for a: 12._____

a) pressure switch.   c) proximity switch.

b) temperature switch.   d) level switch.

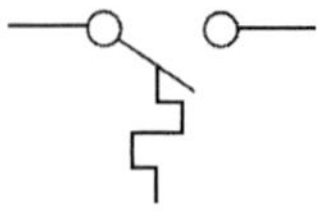

Figure 6-8 Symbol for question 13.

13. Figure 6-8 represents the symbol for a: 13._____

a) proximity switch.   c) limit switch.

b) temperature switch.   d) level switch.

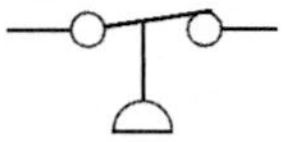

Figure 6-9 Symbol for question 14.

14. Figure 6-9 represents the symbol for a: 14._____

a) pressure switch.   c) limit switch.

b) temperature switch.   d) proximity switch.

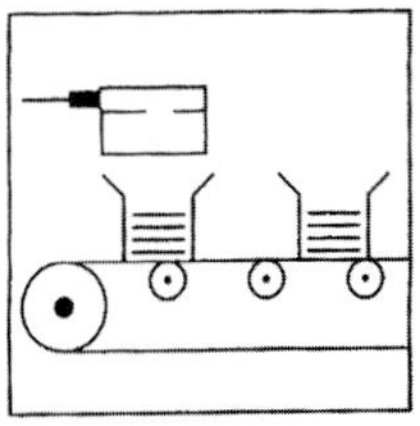

Figure 6-10 Symbol for question 15.

15. Figure 6-10 illustrates a typical application for a: 15._____

a) level switch.   c) proximity switch.

b) temperature switch.   d) limit switch.

Figure 6-11 Symbol for question 16.

16. Figure 6-11 represents the symbol for a: 16._____

a) heater.   c) solenoid.

b) horn.   d) motor.

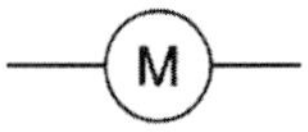

 **Figure 6-12 Symbol for question 17.**

17. Figure 6-12 represents the symbol for a(n): 17._____

a) solenoid valve.   c) overload relay coil.

b) motor starter coil.   d) overload relay contact.

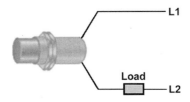 **Figure 6-13 Symbol for question 18.**

18. Figure 6-13 represents the symbol for a: 18._____

a) light sensor.   c) pressure sensor.

b) heat sensor.   d) proximity sensor.

Figure 6-14 Circuit for question 19.

19. The circuit shown in Figure 6-14 is that of a: 19._____

a) one-wire proximity sensor.   c) three-wire proximity sensor.

b) two-wire proximity sensor.   d) four-wire proximity sensor.

20. Most proximity switches come equipped with an LED status 20._____
indicator to:

a) indicate that power is being applied to the switch.

b) indicate that the target is within sensing range.

c) indicate a blown fuse within the switch.

d) verify the output switching action.

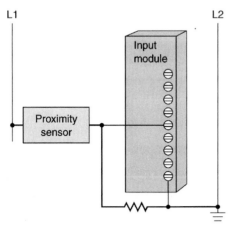

Figure 6-15 Circuit for question 21.

21. The resistor shown in the circuit of Figure 6-15: 21._____

a) is called a bleeder resistor.

b) is used to used to power the sensor continuously.

c) allows enough current for the sensor to operate but not enough to turn on the input.

d) all of these.

22. Capacitor proximity sensors: 22._____

a) are actuated by both conductive and nonconductive materials.

b) are actuated by conductive materials only.

c) produce an electromagnetic field.

d) have a long sensing range.

23. An inductive proximity sensor is actuated by: 23._____

a) a metal object.   c) a light beam.

b) a nonconductive material.   d) any of these.

24. The most common actuator for a reed switch is: 24._____

a) a permanent or electromagnet.   c) application of pressure.

b) a light beam.   d) a heat source.

25. A(n) _____ converts light energy directly into electric energy.   25._____

a) LED   c) solar cell

b) phototransistor   d) photoconductive cell

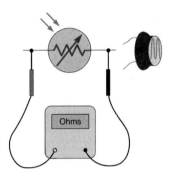

Figure 6-16 Sensor for question 26.

26. The light sensor shown in Figure 6-16 would be classified as a: 26._____

a) solar cell.   c) photovoltaic cell.

b) photoconductive cell.   d) all of these.

27. The light source used in most industrial photoelectric sensors is a(n):   27._____

a) LED.   c) photovoltaic cell.

b) phototransistor.   d) miniature incandescent lamp.

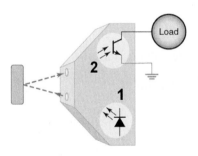

Figure 6-17 Sensor for question 28.

28. For the photoelectric sensor shown in Figure 6-17, part 1 is the ____   28._____
and part 2 is the____.

a) input, output   c) transmitter, receiver

b) primary, secondary   d) high side, low side

29. Fiber optic sensors: 29._____

a) are not affected by electrical interference.

b) carry light signal signals.

c) use a flexible cable containing tiny fibers.

d) all of these.

30. Bar code scanners are used primarily for: 30._____

a) data collection.   c) pressure measurement.

b) temperature measurement.   d) flow measurement.

31. A(n) ____ operates by sending sound waves toward a target and 31._____
measuring the time it takes for the pulses to bounce back.

a) pressure sensor   c) ultrasonic sensor

b) bar code scanner   d) flowmeter

32. The force applied to a strain gauge causes it to bend and change its: 32._____

a) temperature.   c) voltage.

b) resistance.   d) current.

33. A thermocouple, when heated: 33._____

a) produces a small DC voltage.   c) increases its resistance value.

b) produces a small AC voltage.   d) decreases its resistance value.

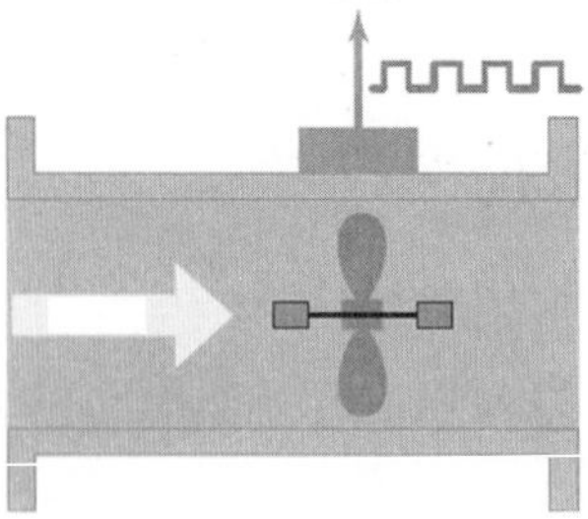

Figure 6-18 Sensor for question 34.

34. The sensor illustrated in Figure 6-18 is a: 34._____

a) thermocouple.   c) strain gauge load cell.

b) turbine flow meter.   d) retroreflective sensor.

35. A tachometer normally refers to a(n) _____ used for speed measurement.   35._____

a) load cell   c) ultrasonic sensor

b) capacitive proximity sensor   d) small generator

36. An encoder is used to: 36._____

a) encode signals from a scanner.   c) change DC to AC.

b) decode signals from a scanner.   d) convert motion into a digital signal.

37. An actuator is a device that: 37._____

a) changes AC to DC.

b) changes DC to AC.

c) converts an electrical signal into mechanical movement.

d) converts mechanical movement into an electrical signal.

38. Solenoid valves are available to control: 38._____

a) oil flow.   c) water flow.

b) air flow.   d) all of these.

39. A(n) _____ converts electrical pulses applied to it into discrete rotor 39._____
movements.

a) tachometer   c) stepper motor

b) solenoid   d) electronic magnetic flowmeter

40. All servo motors: 40._____

a) operate in the closed-loop mode. c) operate without feedback.

b) operate in the open-loop mode.   d) both b and c.

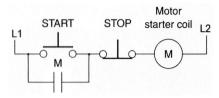

Figure 6-19 Circuit for question 41.

41. The purpose of the M contact shown in the circuit of Figure 6-19 is to: 41._____

a) open the circuit in the case of a motor overload.

b) start the motor from a remote location.

c) keep the starter coil energized when the start button is released.

d) keep the starter coil deenergized when the start button is released.

42. The electromagnetic latching relay: 42._____

a) is used to lock in a condition.

b) uses a latch coil to set and hold the relay in the latched position.

c) uses an unlatch coil to disengage the mechanical latch.

d) all of these.

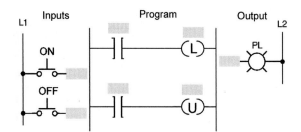

Figure 6-20 Program for question 43.

43. For the programmed latching operation shown in Figure 6-20, which   43._____
two instructions must have the same address?

a) ON and OFF inputs   c) ON input and latch output

b) Latch and unlatch outputs   d) OFF input and unlatch output

44. The Output Latch (OTL) instruction: 44._____

   a) can turn an output on, but it cannot turn the output off.

   b) is used only to turn a bit on and latch it on.

   c) allows the output to remain on even if the rung changes to false.

   d) all of these.

45. Programmed latching circuits are retentive. This means that if power is   45._____
   interrupted the output will _____ when power is returned.

   a) switch to the ON state

   b) switch to the OFF state

   c) remain in its original ON or OFF state

   d) flash ON and OFF

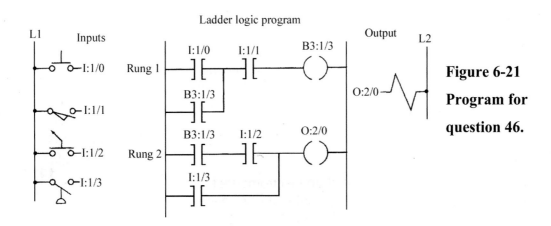

**Figure 6-21
Program for
question 46.**

46-1. For the program of Figure 6-21, what is the address of the instruction   46-1._____
associated with the pressure switch?

a) I:1/0   c) I:1/1

b) I:1/2   d) I:1/3

46-2. What type of the instruction is associated with the selector   46-2._____
switch?

a) OTE   c) XIC

b) OTD   d) XIO

46-3. Rung 1 will be true whenever the: 46-3.____

a) pushbutton and limit switch are closed.

b) selector switch and pressure switch are closed.

c) selector switch and pushbutton are closed.

d) pressure switch and pushbutton are closed.

46-4. Rung 2 will be true whenever: 46-4.____

a) the pressure switch is closed.

b) the pushbutton is closed.

c) the selector switch is closed and rung 1 is true.

d) either a or c.

46-5. The instruction at address B3:1/3 is associated with: 46-5.____

a) an internal relay coil.   c) an external output device.

b) an external input device.   d) the solenoid.

46-6. If the XIC instruction at address I:1/3 is true: 46-6.____

a) output B3:1/3 will also be true.   c) input I:1/2 will also be true.

b) output O:2/0 will also be true.   d) both a and b.

46-7. Assume that an NC limit switch is substituted for the NO limit switch.   46-7.____

For the circuit to operate in the same manner as before:

a) the wires to the limit switch would have to be reversed.

b) the address of the limit switch would have to be changed to I:1/4.

c) the instruction representing the limit switch would have to be changed to an XIO.

d) both b and c.

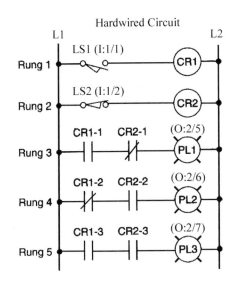

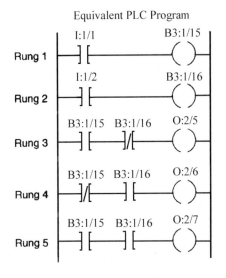

Figure 6-22 Program for question 47.

47-1. With reference to the hardwired circuit and equivalent PLC program of Figure 6-22, when LS2 is actuated and LS1 is not:

a) PLC input I:1/2 is true.

b) PLC programmed rung 2 is false.

c) pilot light PL2 is ON.

d) both a and c.

47-1.____

47-2. When LS1 is actuated and LS2 is not:

a) pilot light PL1 (O:2/5) is ON.

b) pilot light PL2 (O:2/6) is ON.

c) pilot light PL3 (O:2/7) is ON.

d) all of these.

47-2.____

47-3. When both LS1 and LS2 are not actuated:

a) PLC program rung 1 is false.

b) PLC program rung 2 is true.

c) pilot light PL2 (O:2/6) is ON.

d) all of these.

47-3.____

47-4. When both LS1 and LS2 are actuated:

a) PLC program rung 2 is true.

b) PLC program rung 3 is true.

c) PLC program rung 4 is true.

d) PLC program rung 5 is true.

47-4.____

TEST 6.2

Place the answers to the following questions in the answer column at the right.

1. An electromechanical relay uses electromagnetism to operate contacts. (True or False)

1._____

2. When current flows through the coil of a relay, the coil is said to be _____.

2._____

3. A normally closed (NC) relay contact is closed when current flows through the coil. (True or False)

3._____

4. A relay usually will have only one coil but a number of different contacts. (True or False)

4._____

5. Each contact of a relay is usually drawn as it would appear with the coil _____ .

5._____

6. In general, control relay contacts are designed to handle higher currents than contactors. (True or False)

6._____

7. In general, PLC output modules are designed to switch high current loads directly. (True or False)

7._____

8. A motor starter is made up of a contactor with a(n) _____ relay physically attached to it.

8._____

9. In a magnetic motor starter, the control circuit is required to handle the full load current of the motor. (True or False)

9._____

10. In a motor starter, a(n) _____ relay is provided to protect the motor against excessive current.

10._____

11. A(n) _____ operated switch is controlled by hand.

11._____

12. A selector switch is rotated to open and close contacts. (True or False)

12._____

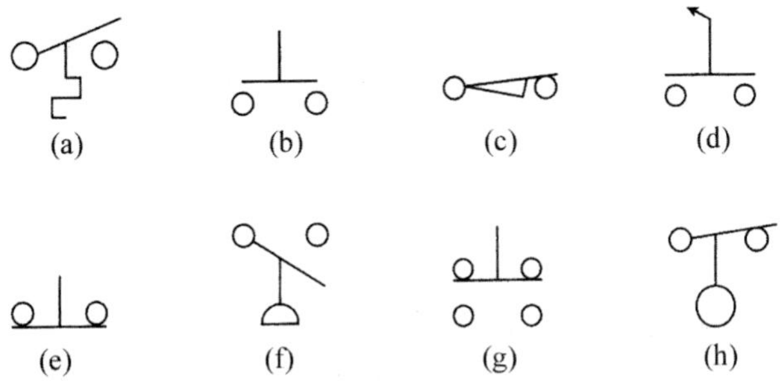

Figure 6-23 Program for question 13.

13. Identify the symbols for the input devices shown in Figure 6-23 from the following list: NO pushbutton, NC pushbutton, break-make pushbutton, selector switch, NC limit switch, NO temperature switch, NO pressure switch, and NC level switch.

13a._____

13b._____

13c._____

13d._____

13e._____

13f._____

13g._____

13h._____

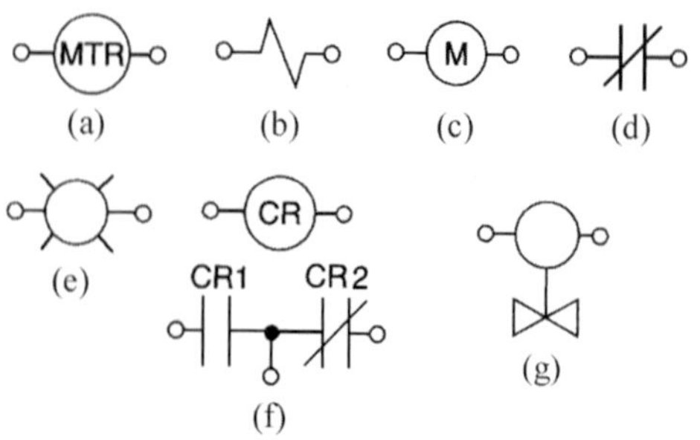

Figure 6-24 Program for question 14.

14. Identify the symbols for the output devices shown in Figure 6-24 from the following list: pilot light, control relay, motor starter coil, OL relay contact, solenoid, solenoid valve, and motor.

14a._____

14b._____

14c._____

14d._____

14e._____

14f._____

14g._____

15. Proximity switches usually sense the presence or absence of a target   15._____
by means of physical contact. (True or False)

16. The small difference between the ON and OFF point of a proximity   16._____
sensor is known as _____.

17. A small current flows through a solid-state proximity sensor even when   17._____
the output is turned off. (True or False)

18. An inductive proximity sensor is actuated by conductive and   18._____
nonconductive materials. (True or False)

19. LED light sources used in photoelectric sensors normally emit a continuous beam of light. (True or False)

19._____

20. A through-beam photoelectric sensor is used to detect the light beam reflected from a target. (True or False)

20._____

21. Bar code _____ are used by PLCs to read bar codes on products.

21._____

22. Changes to dual-in-line (DIP) package switch settings are made mainly during installation. (True or False)

22._____

23. For a thermocouple to generate a voltage, a temperature difference must exist between the (a) _____ and (b) _____ junctions.

23a._____
23b._____

24. Ultrasonic sensors operate using high-frequency light waves. (True or False)

24._____

25. Weight strain gauges operate by detecting small changes in temperature. (True or False)

25._____

26. Reed switches are activated by magnetism. (True or False)

26._____

27. The usual approach to flow measurement is to convert the _____ energy that the fluid has into some other measurable form.

27._____

28. A(n) _____ generator converts rotational speed into a voltage signal used for speed measurement.

28._____

29. Encoders are used in applications where positions have to be precisely determined. (True or False)

29._____

30. Fiber optic sensor systems are completely immune to all forms of electrical interference. (True or False)

30._____

31. A solenoid is made up of a(n) (a) _____ with (b) _____ iron core.

31a._____

31b._____

32. A one-degree-per-step stepper motor requires _____ pulses to move through one revolution.

32._____

33. All servo motors operate in open loop. (True or False)

33._____

34. A _____ circuit is a method of maintaining current flow after a momentary switch has been pressed and released.

34._____

35. Latching relays are used when it is necessary for contacts to stay open and/or closed, even though the coil is momentarily energized. (True or False)

35._____

36. The electromagnetic latching relay function can be programmed on a PLC using the (a) _____ and (b) _____ output instructions.

36a._____

36b._____

37. The programmed latching relay instruction is retentive; that is, if the relay is latched, it will unlatch if power is lost and then restored. (True or False)

37._____

38. A(n) _____ control process is required for processes that require certain operations be performed in a specific order.

38._____

39. A(n) _____ control process is required for processes that require certain operations be performed without regard to the order in which they are performed.

39._____

40. There is more than one correct way to implement the ladder logic for a given control process. (True or False)

40._____

Programming Assignments

This section presents several common programming conversion applications designed to give you, the student, a feel for the potential of the ladder logic programming language. The instructions used are intended to be generic in nature and, as such, will require some conversion for the particular PLC model you are using. The use of a prewired PLC input/output control panel is recommended to simulate the operation of these circuits.

1) The hardwired circuit shown in Figure 6-25 can be programmed to show how contacts of a programmed output relay can be examined for an ON or OFF condition as many times as you like. Prepare an I/O connection diagram and ladder logic program for the circuit. Enter the program into the PLC, and prove its operation.

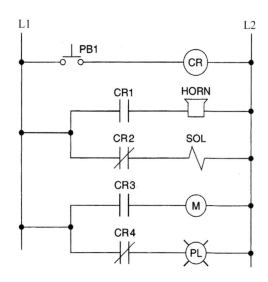

Figure 6-25 Hardwired circuit for assignment 1.

2) The hardwired circuit shown in Figure 6-26 can be programmed to show how a single-pole input device can be programmed as a double-pole input device. Prepare an I/O connection diagram and ladder logic program for the circuit using only the single set of NO contacts of the pressure switch. Enter the program into the PLC, and prove its operation.

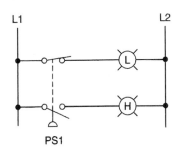

Figure 6-26 Hardwired circuit for assignment 2.

3) The hardwired start/stop motor control circuit shown in Figure 6-27 can be programmed using a PLC. Prepare an I/O connection diagram and ladder logic program for the circuit. Enter the program into the PLC, and prove its operation.

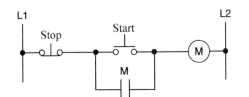

Figure 6-27 Hardwired circuit for assignment 3.

4) The hardwired multiple start/stop motor control circuit shown in Figure 6-28 can be programmed using a PLC. Prepare an I/O connection diagram and ladder logic program for the circuit. Enter the program into the PLC, and prove its operation.

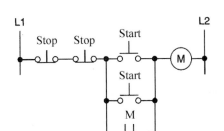

Figure 6-28 Hardwired circuit for assignment 4.

5) The hardwired manual/automatic circuit shown in Figure 6-29 can be programmed using a PLC. The operation of the process is summarized as follows:

- The pump is started by pressing the start button.

- When the selector switch is in the manual position, the solenoid valve is energized at all times.

- When the selector switch is in the automatic position, the solenoid valve is energized only when the pressure switch is closed.

Prepare an I/O connection diagram and ladder logic program for the circuit. Enter the program into the PLC, and prove its operation.

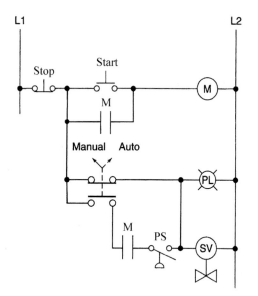

Figure 6-29 Hardwired circuit for assignment 5.

6) The hardwired reversing motor starter circuit shown in Figure 6-30 can be programmed using a PLC. Prepare an I/O connection diagram and ladder logic program for the circuit. Enter the program into the PLC, and prove its operation.

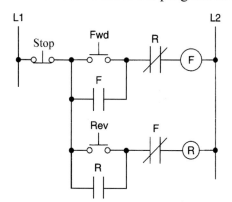

Figure 6-30 Hardwired circuit for assignment 6.

7) The hardwired electric door opener circuit in Figure 6-31 can be programmed using a PLC. Prepare an I/O connection diagram and ladder logic program for the circuit. Enter the program into the PLC, and prove its operation.

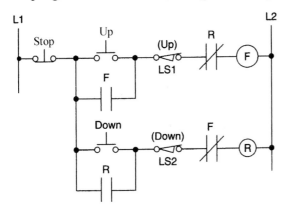

Figure 6-31 Hardwired circuit for assignment 7.

8) The hardwired reciprocating machine process circuit shown in Figure 6-32 can be programmed using a PLC. The operation of the process is summarized as follows:

- The work piece starts on the left and moves to the right when the start button is pressed.
- When the piece reaches the rightmost limit, the drive motor reverses and brings the piece back to the leftmost position.
- This operation is then repeated.
- The forward and reverse pushbuttons provide a means of starting the motor in either forward or reverse so that the limit switches can take over automatic control.

Prepare an I/O connection diagram and ladder logic program for the circuit. Enter the program into the PLC, and prove its operation.

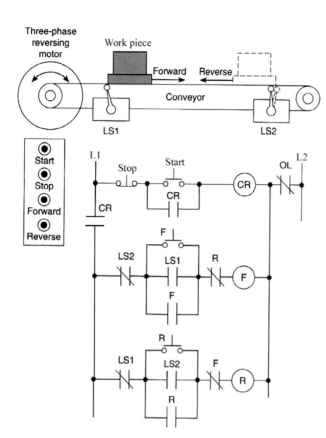

Figure 6-32 Hardwired circuit for assignment 8.

9) The hardwired interlocking circuit of two motors shown in Figure 6-33 can be programmed using a PLC. The application requires that motor No. 2 cannot be started unless motor No. 1 is running. Prepare an I/O connection diagram and ladder logic program for the circuit. Enter the program into the PLC, and prove its operation.

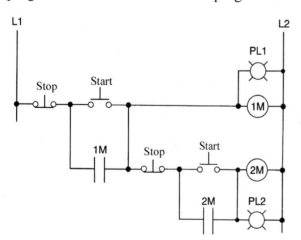

Figure 6-33 Hardwired circuit for assignment 9.

10) The hardwired pushbutton interlock circuit of Figure 6-34 can be programmed using a PLC. The output is energized if button A or button B is pressed but not if both are pressed. Prepare an I/O connection diagram and ladder logic program for the circuit. Enter the program into the PLC, and prove its operation.

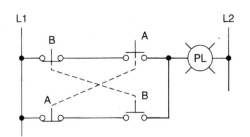

Figure 6-34 Hardwired circuit for assignment 10.

11a) The ladder logic program shown in Figure 6-35 is a push-to-start/push-to-stop operation. The operation of the program is summarized as follows:

- A single momentary normally open pushbutton (I:1/0) performs both the start and stop functions.
- The first time the pushbutton is pressed, internal relay instruction B3:1/11 is latched, energizing output O:2/0.
- The second time you press the pushbutton, internal relay instruction B3:1/12 unlatches instruction B3:1/11, deenergizing output O:2/0.
- Internal relay instruction B3:1/10 prevents any interaction between instructions B3:1/12 and B3:1/11.

Prepare an I/O connection diagram and ladder logic program for the circuit. Use addresses that apply to your PLC. Enter the program into the PLC, and prove its operation.

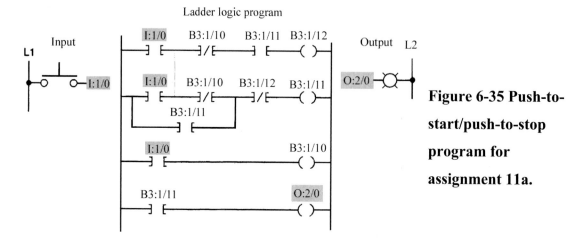

Figure 6-35 Push-to-start/push-to-stop program for assignment 11a.

11b) The program in Figure 6-35 is to be used to control the light ON and OFF from four remote locations. Assuming that one NO pushbutton is used at each location, prepare an I/O connection diagram and ladder logic program for the circuit. Enter the program into the PLC, and prove its operation.

12a) Prepare an I/O connection diagram and a ladder logic program for the Latch/Unlatch program shown in Figure 6-36. Enter the program into the PLC and prove its operation.

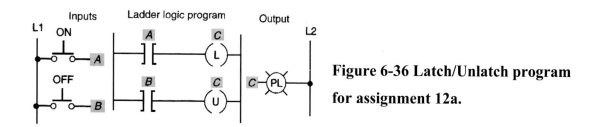

Figure 6-36 Latch/Unlatch program for assignment 12a.

12b) Operate the circuit with both the ON (Latch) and OFF (Unlatch) pushbuttons pressed. Make note of the status of the light (ON or OFF) at all times?. With reference to the way the controller executes the program, explain why the light appears to be ON or OFF at all times.

12c) Repeat 12b with the order of the program changed as shown in the ladder logic program in Figure 6-37.

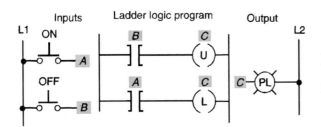

Inputs   Ladder logic program   Output

Figure 6-37 Latch/Unlatch program for assignment 12c.

13) Design two different PLC programs of the hardwired motor start/stop control circuit shown in Figure 6-38. One design is to be a seal-in type, start/stop control, and the other is to be a latch/unlatch type start/stop control. Incorporate both designs into a single program. Enter the program into the PLC, and prove its operation. With both outputs initially ON, simulate a power failure to the PLC system. When power was restored, what happened to the outputs? When would you use one circuit over the other?

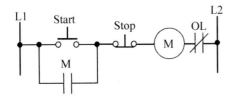

Figure 6-38 Hardwired control circuit for assignment 13.

14) Design a PLC program that will perform the following tasks:

a) When switch 1 is closed, three lights will come on.

b) If switch 2 is then closed, two of the lights will drop out, leaving one light on.

c) If switch 3 is then closed (switches 1, 2, and 3 all closed at this point), one of the lights that dropped out in operating state *b* above will come on, thus showing two lights on.

d) Closing switch 4 turns off any of the three lights that happen to be on.

Enter the program into the PLC, and prove its operation.

15) There are four normally open input sensors to an annunciator system that switches the output alarm ON if some operational malfunction occurs. Design a program that operates the alarm system as follows:

- If any one input is closed, nothing happens.

- If any two inputs are closed, a green pilot light goes on.

- If any three inputs are closed, a yellow pilot light goes on.

- If all four inputs are closed, a red pilot light goes on.

Enter the program into the PLC, and prove its operation.

16) The program of Figure 6-39 is described in the text and is used to control the level of water in a storage tank. Prepare an I/O connection diagram and ladder logic program for the process. Use addresses that apply to your PLC. Enter the program into the PLC, and prove its operation.

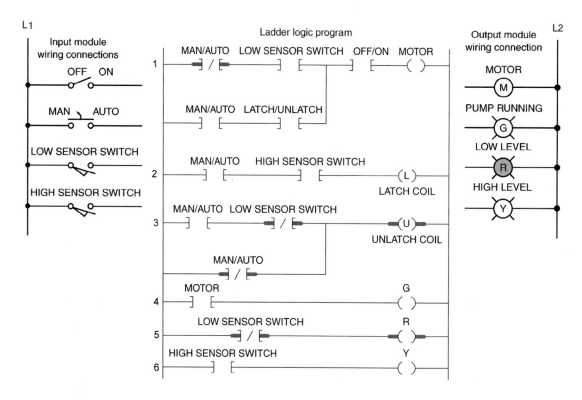

Figure 6-39 Water level control program for assignment 16.

17) The sequential process program of Figure 6-40 is described in the text. Prepare an I/O connection diagram and ladder logic program for the process. Use addresses that apply to your PLC. Enter the program into the PLC, and prove its operation.

Ladder logic program

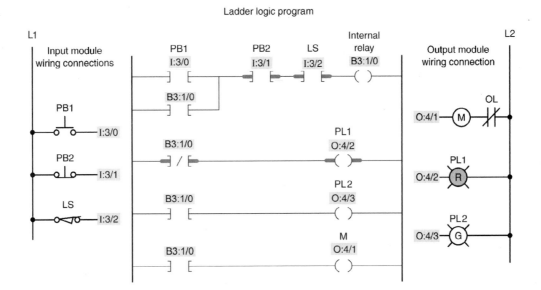

Figure 6-40 Sequential process program for assignment 17.

18) The motor jog program of Figure 6-41 is described in the text. Prepare an I/O connection diagram and ladder logic program for the process. Use addresses that apply to your PLC. Enter the program into the PLC, and prove its operation.

Ladder logic program

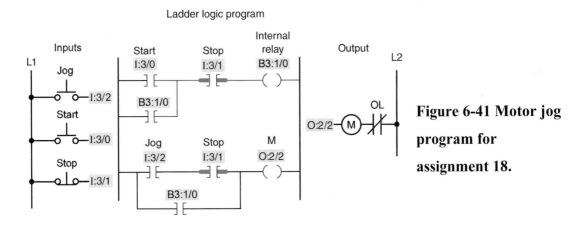

Figure 6-41 Motor jog program for assignment 18.

19) The drilling process program of Figure 6-42 is described in the text. Prepare an I/O connection diagram and ladder logic program for the process. Use addresses that apply to your PLC. Enter the program into the PLC, and prove its operation.

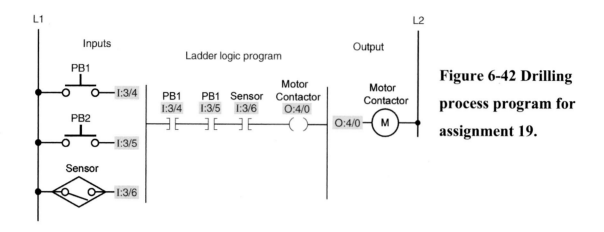

Figure 6-42 Drilling process program for assignment 19.

20) The motorized door program of Figure 6-43 is described in the text. Prepare an I/O connection diagram and ladder logic program for the process. Use addresses that apply to your PLC. Enter the program into the PLC, and prove its operation.

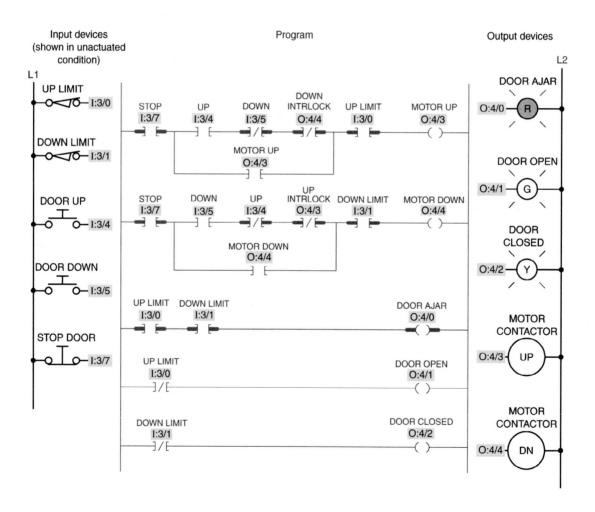

Figure 6-43 Motorized door program for assignment 20.

21) The continuous filling program of Figure 6-44 is described in the text. Prepare an I/O connection diagram and ladder logic program for the process. Use addresses that apply to your PLC. Enter the program into the PLC, and prove its operation.

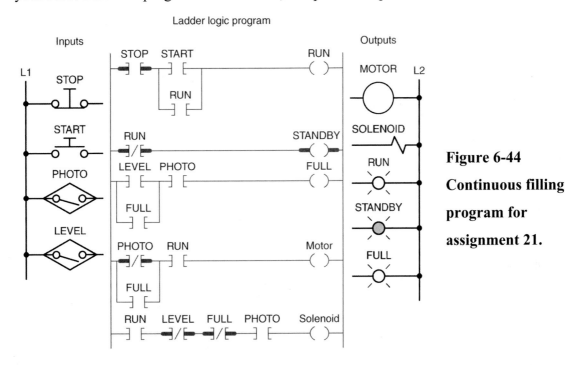

Figure 6-44
Continuous filling program for assignment 21.

22) Design a ladder logic program that will cause output pilot light PL to be ON when selector switch SS is closed, pushbutton PB is open, and limit switch LS is open. Enter the program into the PLC, and prove its operation.

23) Design a ladder logic program that will cause a solenoid, SOL, to be energized when limit switch LS is closed and pressure switch PS is open. Enter the program into the PLC, and prove its operation.

24) Design a ladder logic program that will cause output pilot light PL to be latched when pushbutton PB1 is closed, and unlatched when either pushbutton PB2 or pushbutton PB3 is closed. Also, do not allow the unlatch to go true when the latch rung is true, nor allow the latch rung to go true when the unlatch rung is true. Enter the program into the PLC, and prove its operation.

25) Design a ladder logic program that will cause pilot light PL to be ON when pushbutton PB is closed and either limit switch LS1 or LS2 is closed. Enter the program into the PLC, and prove its operation.

26) Design a ladder logic program that will cause pilot light PL to be ON when pushbutton PB1 is open, pushbutton PB2 is closed, and either LS1 is open or limit switch LS2 is closed. Enter the program into the PLC, and prove its operation

CHAPTER 7   Programming Timers

TEST 7.1

Choose the letter that best completes the statement.

1. Certain contacts of a mechanical timing relay are designed to operate   1._____
at a preset time interval:
a) after the coil is energized.   c) after power is applied to the circuit.
b) after the coil is deenergized.   d) either a or b.

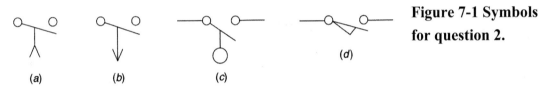

(a)   (b)   (c)   (d)

**Figure 7-1 Symbols
for question 2.**

2. Which of the symbols shown in Figure 7-1 represents an on-delay timed   2._____
relay contact?

     Figure 7-2 Contact symbol for question 3.

3. The relay contact drawn in Figure 7-2 is designed to operate so that: 3._____
a) when the relay coil is energized, there is a time delay in closing.
b) when the relay coil is energized, there is a time delay in opening.
c) when the relay coil is deenergized, there is a time delay before the contact opens.
d) when the relay coil is deenergized, there is a time delay before the contact closes.

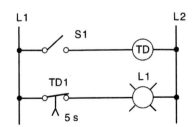   **Figure 7-3 Hardwired circuit for question 4.**

4. In the hardwired circuit of Figure 7-3, the light will stay on: 4._____
a) as long as S1 is closed.   c) for 5 s after coil TD is deenergized.
b) for 5 s after coil TD is energized.   d) both a and c.

5. Which one of the following timer parameters determines the time duration for the timing circuit?

5._____

a) Accumulated time   c) Timer address

b) Preset time   d) Time base

6. Which one of the following timer parameters represents the value that increments as the timer is timing?

6._____

a) Accumulated time   c) Timer address

b) Preset time   d) Time base

7. Which one of the following timer parameters determines the accuracy of the timer?

7._____

a) Accumulated time   c) Timer address

b) Preset time   d) Time base

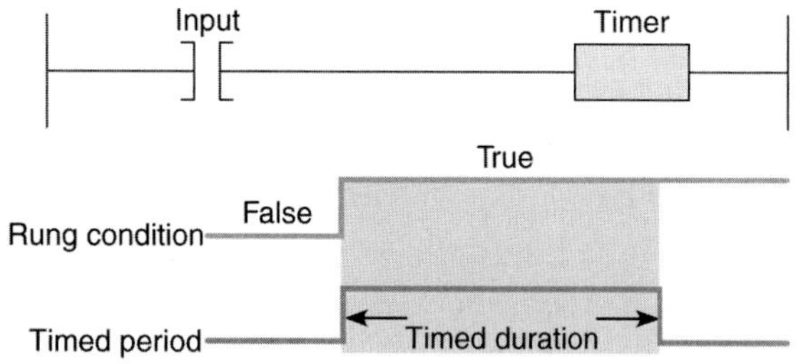

Figure 7-4 Timer program for question 8.

8-1 The timer shown in Figure 7-4 would be classified as a(n):

8-1._____

a) on-delay timer.   c) normally open timer.

b) off-delay timer.   d) normally closed timer.

8-2. The timing commences when:

8-2._____

a) the input instruction is true.   c) power is applied.

b) the input instruction is false.   d) power is removed.

9. The timer file for SLC 500 controllers is:

9._____

a) T1.   c) T3.

b) T2.   d) T4.

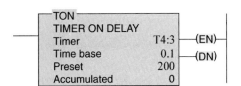

Figure 7-5 Timer instruction for question 10.

10-1. For the on-delay timer instruction shown in Figure 7-5, the timer number is:

10-1.____

a) 0.

c) T4:3.

b) 200.

d) 0.1.

10-2. The on-delay timed period would be:

10-2.____

a) 3 seconds.

c) 20 seconds.

b) 4 seconds.

d) 200 seconds.

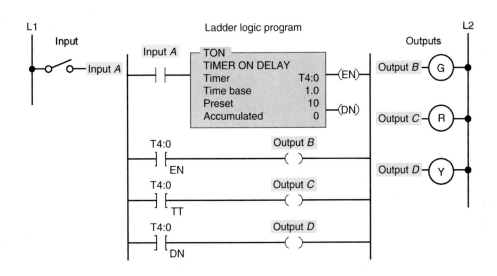

Figure 7-6 Timer program for question 11.

11-1. For the on-delay timer program shown in Figure 7-6, output *B* is switched ON when:

11-1.____

a) power is applied.

b) input *A* is closed.

c) the timer is accumulating time.

d) the accumulated value equals the preset value.

11-2. Output *C* is switched ON when:　　　　　　　　　　　　　　　11-2.____

a) power is applied.

b) input *A* is closed.

(c) the timer is accumulating time.

d) the accumulated value equals the preset value.

11-3. Output *D* is switched ON when:　　　　　　　　　　　　　　　11-3.____

a) power is applied.

b) input *A* is closed.

c) the timer is accumulating time.

(d) the accumulated value equals the preset value.

11-4. The timer accumulated value will reset to zero whenever:　　　11-4.____

(a) input *A* switch is opened.　　　　　(c) the DN bit is set to 1.

b) input *A* switch is closed.　　　　　　d) the EN bit is set to 1.

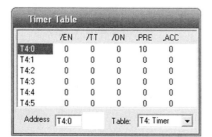

Timer Table

	/EN	/TT	/DN	.PRE	.ACC
T4:0	0	0	0	10	0
T4:1	0	0	0	0	0
T4:2	0	0	0	0	0
T4:3	0	0	0	0	0
T4:4	0	0	0	0	0
T4:5	0	0	0	0	0

Address T4:0　　Table: T4: Timer

Figure 7-7 Timer table for question 12.

12. For the timer table shown in Figure 7-7, bit level addressing is used for:　　12._____

a) EN, TT, PRE, and ACC.

(b) EN, TT, and DN.

c) PRE, ACC, TT, and EN.

d) PRE and ACC.

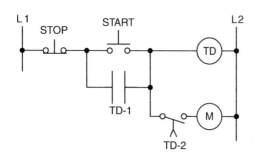

Figure 7-8 Hardwired timer for question 13.

13. For the hardwired timer circuit of Figure 7-8, contact TD-1 is the _____   13._____
contact and TD-2 is the _____ contact.

a) ON, OFF   c) instantaneous, timed

b) OFF, ON   d) timed, instantaneous

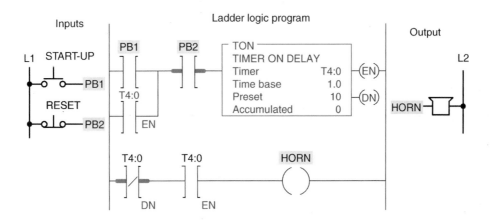

Figure 7-9 Programmed timer for question 14.

14-1. For the programmed timer circuit of Figure 7-9, the _____ bit of the   14-1._____
timer functions similar to an instantaneous contact.

a) DN   c) PB1

b) EN   d) PB2

14-2. The _____ bit of the timer functions similar to a timed contact.   14-2._____

a) DN   c) PB1

b) EN   d) PB2

15. The on-delay timer (TON) starts timing when the timer's: 15._____

a) ladder rung switches from false to true.

b) ladder rung switches from true to false.

c) accumulated value equals its preset value.

d) accumulated is greater than its preset value.

16. The off-delay timer (TOF) starts timing when the timer's: 16._____

a) ladder rung switches from false to true.

b) ladder rung switches from true to false.

c) accumulated value equals its preset value.

d) accumulated is greater than its preset value.

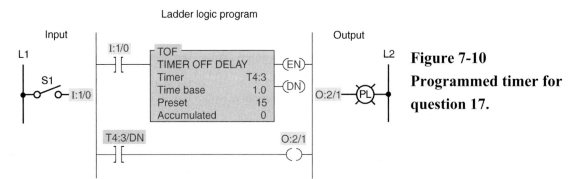

Ladder logic program

Figure 7-10
Programmed timer for
question 17.

17. For the programmed timer circuit of Figure 7-10, the pilot light should   17._____
come on:

a) as soon as the switch is closed.

b) before the switch is closed.

c) for 15 seconds after the switch is opened.

d) both a and c.

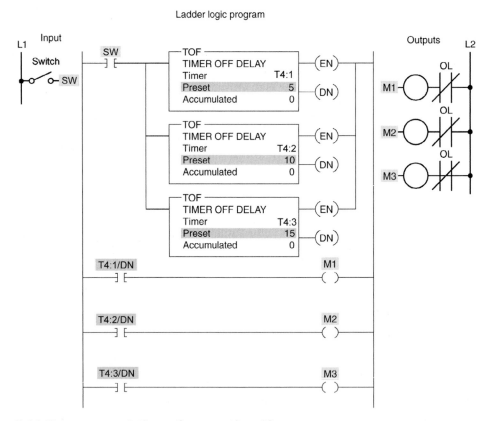

Ladder logic program

Figure 7-11 Programmed timer for question 18.

18-1. For the programmed timer circuit of Figure 7-11, when the switch   18-1.____
is initially closed, motor(s) _____ start(s) immediately.
a) M1   c) M3
b) M2   d) all of these

18-2. Assume the switch is closed for 5 seconds and then opened.   18-2.____
After 12 seconds have elapsed, motor(s) ____ will still be operating.
a) M1, M2, and M3   c) M3
b) M2 and M3   d) none of these

19. The main difference between a TON and TOF timer is that the: 19._____
a) TON timer can maintain its accumulated time on loss of power or logic
continuity.
b) TOF timer can maintain its accumulated time on loss of power or logic
continuity.
c) TOF timer begins timing when logic continuity to the timing rung is lost.
d) TON timer begins timing when logic continuity to the timing rung is lost.

20. The operation of a PLC retentive timer is similar to that of an: 20._____
a) electromagnetic pneumatic timer.   c) off-delay timer.
b) electromechanical motor-driven timer.   d) on-delay timer.

21. The main difference between a PLC retentive and nonretentive timer   21._____
is that the:
a) retentive timer can be programmed for much longer time-delay periods.
b) nonretentive time can be programmed for much longer time-delay periods.
c) retentive timer maintains the current time should power be removed from
the device or when the timer rung goes false.
d) nonretentive timer maintains the current time should power be removed
from the device or when the timer rung goes false.

22. Unlike the TON timer, the RTO timer requires a(n):   22._____
a) timer reset instruction.   c) internal relay instruction.
b) input condition instruction.   d) instantaneous contact instruction.

23. When addressing an RES instruction, it must be addressed to: 23._____

a) a TOF instruction.

b) a TON instruction.

c) any address other than that of the RTO instruction.

d) the same address as that of the RTO instruction.

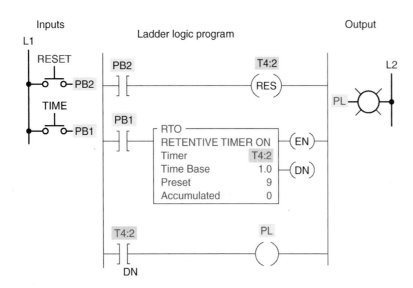

Figure 7-12
Programmed timer for questions 24, 25, and 26.

24-1. The type of timer programmed in Figure 7-12 is a: 24-1.____

a) retentive on-delay.   c) nonretentive off-delay.

b) retentive off-delay.   d) nonretentive on-delay.

24-2. The timer starts timing when: 24-2.____

a) PB1 is closed.   c) both PB1 and PB2 are closed.

b) PB2 is closed.   d) either PB1 or PB2 is closed.

24-3. The timer accumulated time is set to zero any time: 23-4.____

a) PB1 is closed.   c) PB1 is open.

b) PB2 is closed.   d) PB2 is open.

For the timer program of Figure 7-12, assume the following sequence of events:

➢ First PB2 is momentarily pressed closed.

➢ Next PB1 is pressed closed for 5 s and released.

➢ Next PB2 is pressed closed for 4 s and released.

25-1. As a result, the timer-accumulated value at the end of the sequence would be:

25-1._____

a) 5.   c) 9.

b) 4.   d) 0.

25-2. As a result, the output PL would be:

25-2._____

a) on for 4 s and off for 5 s.

b) on for 5 s and off for 4 s.

c) on after the entire sequence has been completed.

d) off after the entire sequence has been completed.

For the timer program of Figure 7-12, assume the following sequence of events:

➢ First PB2 is momentarily pressed closed.
➢ Next PB1 is pressed closed for 3 s and released.
➢ Next PB1 is again pressed closed for 6 s and released.

26-1. As a result, the timer accumulated value at the end of the sequence would be:

26-1._____

a) 3.   c) 9.

b) 6.   d) 0.

26-2. As a result, output PL would be:

26-2._____

a) on for 3 s and off for 6 s.

b) off for 3 s and on for 6 s.

c) on after the entire sequence has been completed.

d) off after the entire sequence has been completed.

27. To reset a retentive timer, the:

27._____

a) AC time must be greater than the PR time.

b) PR time must be greater than the AC time.

c) AC time must equal the PR time.

d) none of these.

28. The interconnecting of timers is commonly called:

28._____

a) grouping.   c) sequencing.

b) programming.   d) cascading.

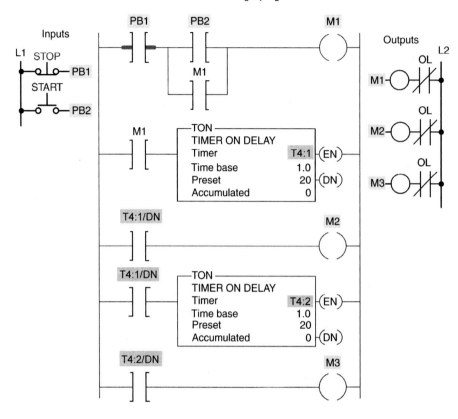

Figure 7-13 Programmed timer for question 29.

29-1. In the timer program of Figure 7-13, the timer T4:1 is energized by actuating: 29-1.____

a) PB1.   c) both PB1 and PB2.

b) PB2.   d) either PB1 or PB2.

29-2. Output M1 is normally energized: 29-2.____

a) as soon as PB1 is actuated.

b) as soon as PB2 is actuated.

c) 10 s after PB1 has been actuated.

d) 40 s after both PB1 and PB2 have been actuated.

29-3. Output M2 is normally energized ____ after output M1 has been energized.   29-3.____

a) 10 s   c) 30 s

b) 20 s   d) 40 s

29-4. Output M3 is normally energized _____ after output M1 has been energized. 29-4.____

a) 10 s   c) 30 s

b) 20 s   d) 40 s

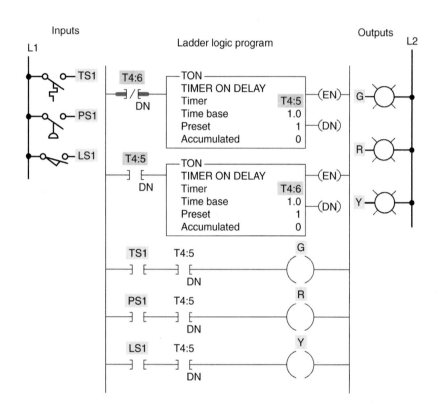

Figure 7-14 Programmed flasher timer for question 30.

30-1. In the annunciator flasher program of Figure 7-14, the two timers are interconnected to form a(n):   30-1.____

a) amplifier circuit.   c) oscillator circuit.

b) rectifier circuit.   d) series-parallel circuit.

30-2. The output of timer T4:5:   30-2.____

a) turns on after a 10-s time delay and remains on.

b) turns off after a 2-s time delay and remains off.

c) turns on after a 3-s time delay and remains on.

d) is pulsed on and off at 1-s intervals.

30-3. When pressure switch PS1 closes, the:

a) green indicating lamp is turned on.

b) red indicating lamp is turned on.

c) yellow indicating lamp is pulsed on and off.

d) red indicating lamp is pulsed on and off.

30-3._____

31. Which instruction can best be used to turn an output coil on or off after the rung has been false for a desired time?

a) RTO

b) TON

c) ONOF

d) TOF

31._____

32. The amount of time for which a timer is programmed is called the:

a) preset.

b) desired time.

c) set point.

d) lapsed time.

32._____

33. When the timing of a device is not reset on a loss of power, the timing is said to be:

a) continuous.

b) holding.

c) retentive.

d) saved.

33._____

34. RES instructions are used with:

a) TOF timers.

b) TON timers.

c) RTO timers.

d) all of these

34._____

TEST 7.2

Place the answers to the following questions in the answer column at the right.

1. Mechanical timing relays are used to ___ the opening or closing of contacts for circuit control.

1._____

2. An off-delay timing relay provides time delay when its coil is ____.

2._____

Figure 7-15 Timed contact for question 3.

3. For the timer relay contact shown in Figure 7-15, when the relay coil is energized, there is a time delay before the contact closes. (True or False)

3._____

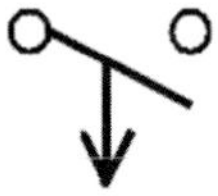

Figure 7-16 Timed contact for question 4.

4. For the timer relay contact shown in Figure 7-16, when the relay coil is deenergized, there is a time delay before the contact opens. (True or False)

4._____

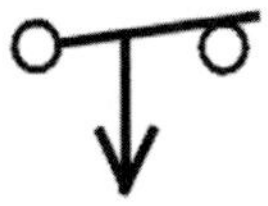

Figure 7-17 Timed contact for question 5.

5. For the timer relay contact shown in Figure 7-17, when the relay coil is deenergized, there is a time delay before the contact closes. (True or False)

5._____

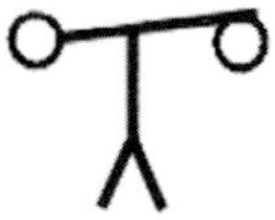

Figure 7-18 Timed contact for question 6.

6. For the timer relay contact shown in Figure 7-18, when the relay coil is energized, there is a time delay before the contact opens. (True or False)

6._____

7. PLC timers are input instructions that provide the same functions as mechanical timing relays. (True or False)

7._____

8. Timer instructions are found on all PLCs manufactured today. (True or False)

8._____

9. Timer instructions may be (a) _____ formatted or (b) _____ formatted.

9a._____
9b._____

10. The parts of a timer instruction include (a) ___, (b) ___, (c) ___, (d) ___, and (e) ____.

10a._____
10b._____
10c._____
10d._____
10e._____

11. The timer output is energized when the (a) _____ time equals the (b) _____ time.

11a._____
11b._____

12. If the preset time of a timer is 100 and the time base is 0.1, the time-delay period would be ____ s.

12._____

13. A(n) _____ timer must be intentionally reset with a separate signal.

13._____

14. The retentive timer reset (RES) instruction is always given the same address as the timer it resets. (True or False)

14._____

15. An alarm is to be switched on whenever a piping system has sustained a cumulative overpressure of 60 s. The most directly applicable timer to use would be the on-delay nonretentive timer. (True or False)

15._____

16. A lamp is to be switched off 10 s after a switch has been switched from its ON to OFF position. The most directly applicable timer to use would be the off-delay nonretentive timer. (True or False)

16._____

17. When a time-delay period longer than the maximum preset time allowed for a single timer is required, the problem can be solved by programming two or more timers together. (True or False)

17._____

18. Normally, the reset input to a timer will override the control input of the timer. (True or False)

18._____

19. A retentive timer must be completely timed out to be reset.   19._____
(True or False)

20. Retentive timers lose the accumulated time every time the rung condition   20._____
becomes false. (True or False)

21. The instantaneous contacts of a timer have no time-delay period   21._____
associated with them. (True or False)

22. What timer instruction (TON, TOF, or RTO) would be best suited for each of the
following control application:

a) Keep track of the total time to make one batch of product, even if the   22a._____
process is halted and then started again.
b) Hold the clamp on for 25 s after the glue is applied.   22b._____
c) Open a valve 27 s after a switch is turned on. If interrupted, the valve   22c._____
should close and the time should reset to 0.
d) Begin timing when the rung is true, and hold the accumulated time   22d._____
when rung logic goes false.

23. The accumulated time of a TOF timer is reset by causing the rung   23._____
to go true momentarily. (True or False)

24. A timer's _____ is the length of time the timer is to time.   24._____

25. A timer's _____ specifies at what rate the timer will increment.   25._____

26. A timer's _____ value is the current elapsed time.   26._____

27. A RES (reset) instruction must be used to zero the accumulated   27._____
value in an RTO timer. (True or False)

28. A timer's delay time equals the value in the ACC multiplied by   28._____
the time base. (True or False)

29. Timers can be retentive or nonretentive. (True or False)   29._____

30. An RTO timer retains the present accumulated value when the rung goes false. (True or False)

30._____

31. A TOF timer starts to accumulate time when the rung becomes true. (True or False)

31._____

32. A TOF timer starts to accumulate time when the rung makes a true to false transition. (True or False)

32._____

Programming Assignments

This section presents several common timing program applications. The instructions used are intended to be generic in nature and, as such, will require some conversion for the particular PLC model you are using. The use of a prewired PLC input/output control panel is recommended to simulate the operation of these circuits.

1) Prepare an I/O connection diagram and ladder logic program for a nonretentive on-delay timer that will turn a light on 5 s after a switch is closed. Enter the program into the PLC, and prove its operation.

2) Prepare an I/O connection diagram and ladder logic program for a nonretentive off-delay timer that will turn a light off 10 s after a switch is opened. Enter the program into the PLC, and prove its operation.

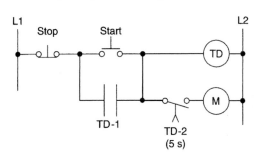

Figure 7-19 Hardwired timer circuit for assignment 3.

3) Prepare an I/O connection diagram and ladder logic program to execute the hardwired timer circuit shown in Figure 7-19. Enter the program into the PLC, and prove its operation.

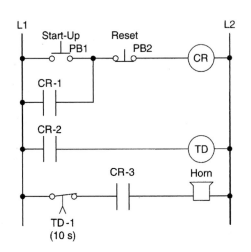

Figure 7-20 Hardwired start-up warning circuit for assignment 4.

4) Prepare an I/O connection diagram and ladder logic program to execute the hardwired start-up warning signal circuit shown in Figure 7-20. Enter the program into the PLC, and prove its operation.

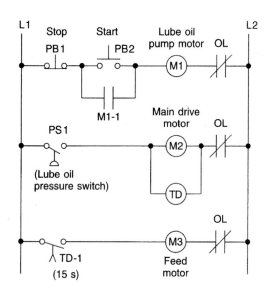

Figure 7-21 Hardwired sequence control circuit for assignment 5.

5) Prepare an I/O connection diagram and ladder logic program to execute the hardwired automatic sequential control system shown in Figure 7-21. Enter the program into the PLC, and prove its operation.

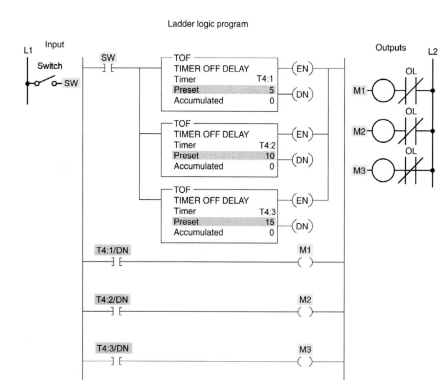

Figure 7-22 Off-delay timer program for assignment 6.

6) The program of Figure 7-22 is described in the text and uses off-delay timers to switch motors off sequentially at 5-second intervals. Prepare an I/O connection diagram and ladder logic program for the process. Use addresses that apply to your PLC. Enter the program into the PLC, and prove its operation.

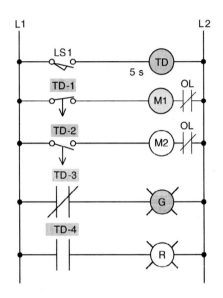

Figure 7-23 Hardwired off-delay timer circuit for assignment 7.

7) Prepare an I/O connection diagram and ladder logic program to execute the hardwired off-delay timer circuit shown in Figure 7-23. Enter the program into the PLC, and prove its operation.

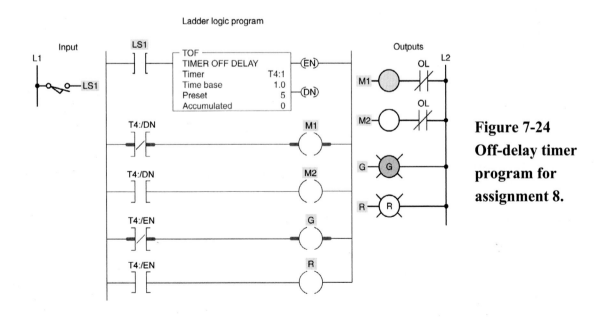

Figure 7-24 Off-delay timer program for assignment 8.

131

8) The program of Figure 7-24 is described in the text and uses an off-delay timer containing both instantaneous and timed contacts. Prepare an I/O connection diagram and ladder logic program for the process. Use addresses that apply to your PLC. Enter the program into the PLC, and prove its operation.

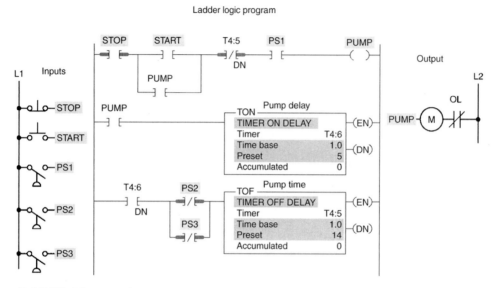

Figure 7-25 Fluid pumping process program for assignment 9.

9) The program of Figure 7-25 is described in the text and uses both on-delay and off-delay timers as part of a fluid pumping process. Prepare an I/O connection diagram and ladder logic program for the process. Use addresses that apply to your PLC. Enter the program into the PLC, and prove its operation.

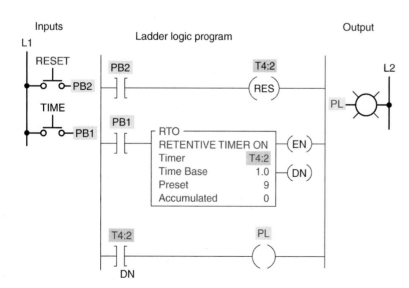

Figure 7-26 Retentive on-delay timer program for assignment 10.

10) Prepare an I/O connection diagram and ladder logic program to execute the retentive on-delay timer program shown in Figure 7-26. Enter the program into the PLC, and prove its operation.

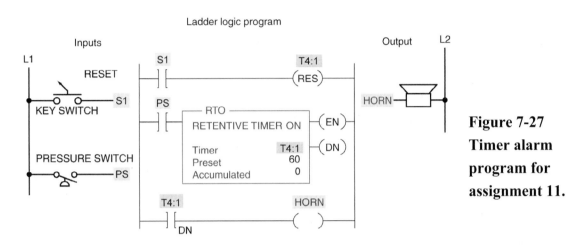

Figure 7-27 Timer alarm program for assignment 11.

11) The program of Figure 7-27 is described in the text and uses a retentive on-delay as part of an alarm program. Prepare an I/O connection diagram and ladder logic program for the process. Use addresses that apply to your PLC. Enter the program into the PLC, and prove its operation.

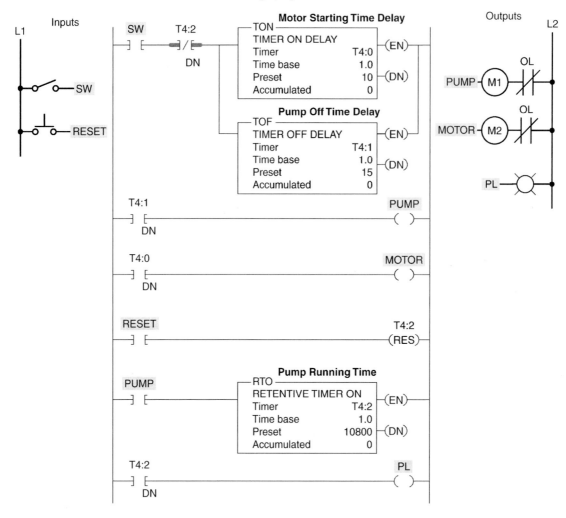

Figure 7-28 Bearing lubrication program for assignment 12.

12) The program of Figure 7-28 is described in the text and uses TON, TOF, and RTO timers as part of a bearing lubrication program. Prepare an I/O connection diagram and ladder logic program for the process. Use addresses that apply to your PLC. Enter the program into the PLC, and prove its operation.

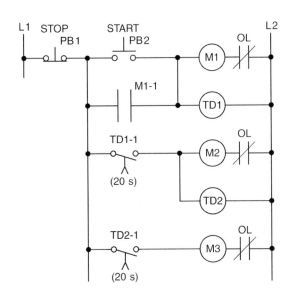

Figure 7-29 Hardwired sequential delayed motor starting circuit for assignment 13.

13) Prepare an I/O connection diagram and ladder logic program to execute the hardwired sequential time delayed motor starting circuit shown in Figure 7-29. Enter the program into the PLC, and prove its operation.

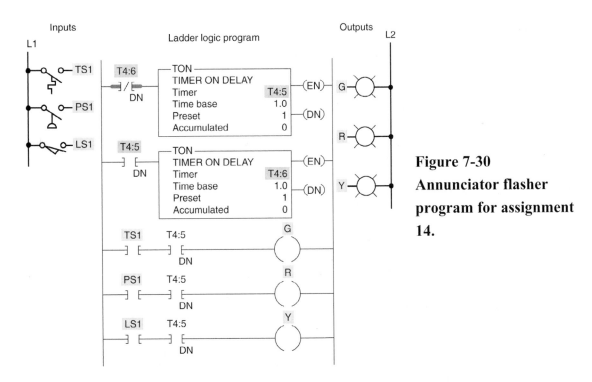

Figure 7-30 Annunciator flasher program for assignment 14.

14) The program of Figure 7-30 is described in the text and uses two interconnected TON timers to form an annunciator flasher program. Prepare an I/O connection diagram and ladder logic program for the process. Use addresses that apply to your PLC. Enter the program into the PLC, and prove its operation.

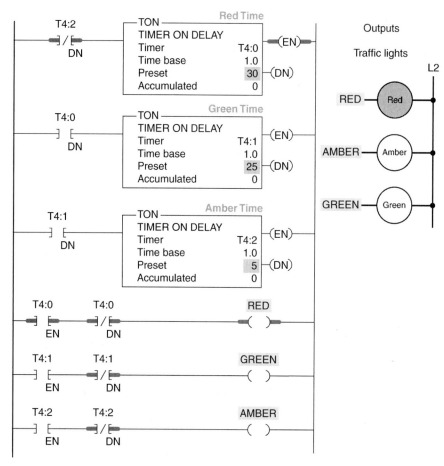

Figure 7-31 Traffic lights program for assignment 15.

15) The program of Figure 7-31 is described in the text and is used for control of traffic lights in one direction. Prepare an I/O connection diagram and ladder logic program for the process. Use addresses that apply to your PLC. Enter the program into the PLC, and prove its operation

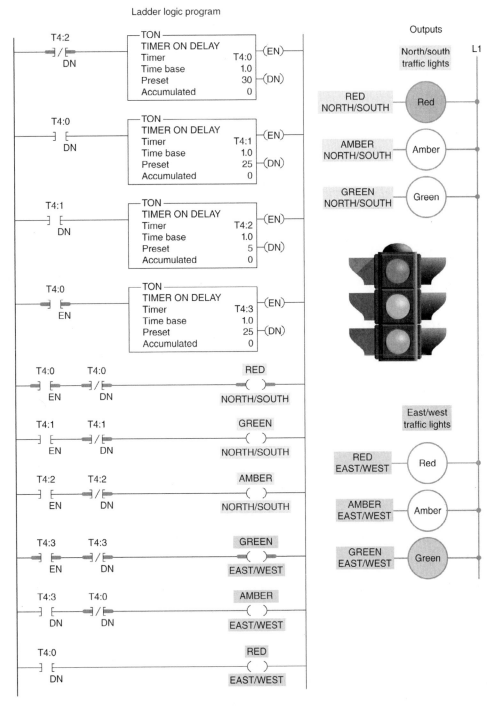

Figure 7-32 Traffic lights program for assignment 16.

16) The program of Figure 7-32 is described in the text and is used for control of traffic lights in two directions. Prepare an I/O connection diagram and ladder logic program for the process. Use addresses that apply to your PLC. Enter the program into the PLC, and prove its operation

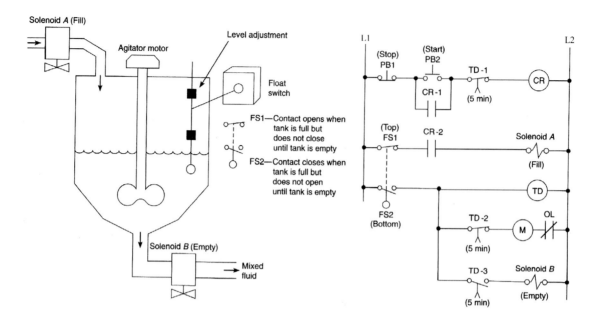

Figure 7-33 Hardwired automatic mixing process for assignment 17.

17) Prepare an I/O connection diagram and ladder logic program for the hardwired automatic mixing process circuit shown in Figure 7-33. Enter the program into the PLC, and prove its operation. The operation of the process can be summarized as follows:

- Process is initiated by pressing the start pushbutton PB2.
- Solenoid A is energized to allow fluid to flow into the tank.
- Fluid flows in until the float switch is activated at the full position.
- Agitator motor is started and operates for 5 min and then stops.
- Solenoid B is energized to empty the tank.
- When the float switch is activated at the empty position, the process stops and is placed in the ready position for the next manual start.

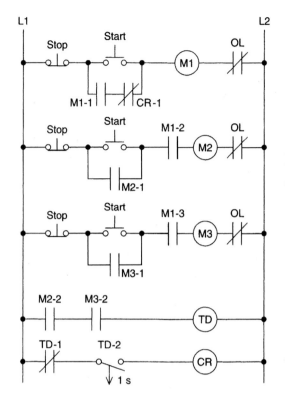

Figure 7-34 Hardwired motor control circuit for assignment 18.

18) Prepare an I/O connection diagram and ladder logic program for the hardwired motor control circuit shown in Figure 7-34. Enter the program into the PLC, and prove its operation. The operation of the process can be summarized as follows:

- The control process consists of three motors: M1, M2, and M3.
- The electrical control system is to be designed so that motor M1 must be running before motor M2 or M3 can be started.
- Each motor has its own start/stop pushbutton station.
- Both motors M2 and M3 can normally be stopped or started without affecting the operation of motor M1.
- However, if all three motors are running, the stopping of any one motor, for any reason, will automatically stop all three motors.

19) Write a PLC ladder logic diagram for a display sign that will sequentially turn on three lights 2 seconds apart and then turn all three lights off and repeat the sequence. Enter the program into the PLC, and prove its operation.

20) Write a PLC program that will turn on pilot light PL1 10 seconds after switch S1 is turned on. Pilot light PL2 will come on 5 seconds after PL1 comes on. Pilot light PL3 will come on 8 seconds after PL2 comes on. Pressing pushbutton PB1 will reset all the timers but only if PL3 is on. Enter the program into the PLC, and prove its operation.

21) When a switch is turned on, PL1 goes on immediately and PL2 goes on 9 seconds later. Opening the switch turns both lights off. Write a program that implements this process. Enter the program into the PLC, and prove its operation.

22) When a switch is turned on, PL1 and PL2 immediately go on. When the switch is turned off, PL1 immediately goes off. PL2 remains on for another 3 seconds and then goes off. Write a program that will implement this process. Enter the program into the PLC, and prove its operation.

23) When a switch is turned on, PL1 and PL2 immediately go on. PL1 turns off after 4 seconds. PL2 remains on until the switch is turned off. Turning the switch off at any time turns both lights off. Write a program that will implement this process. Enter the program into the PLC and prove its operation.

24) A saw, fan, and lube pump all go on when a start button is pressed. Pressing a stop button immediately stops the saw but allows the fan to continue operating. The fan is to run for an additional 5 seconds after shutdown of the saw. If the saw has operated for more than 20 seconds, the fan should remain on until reset by a separate fan reset button. If the saw has operated less than 20 seconds, the lube pump should go off when the saw is turned off. However, if the saw has operated for more than 20 seconds, the lube pump should remain on for an additional 10 seconds after the saw is turned off. Write a program that will implement this process. Enter the program into the PLC, and prove its operation.

CHAPTER 8   Programming Counters

TEST 8.1

Choose the letter that best completes the statement.

1. Programmed counters can:
a) count up.
b) count down.
c) be combined to count up and down.
d) all of these.

1.___D___

2. The counter instruction is found on:
a) all PLCs.
b) small-size PLCs.
c) medium-size PLCs.
d) large-size PLCs.

2.___A___

3. The PLC counter instruction is similar to the:
a) internal relay instruction.
b) transitional contact instruction.
c) relay coil and contact instruction.
d) timer instruction.

3.___D___

4. The output of a PLC counter is energized when the:
a) accumulated count equals the preset count.
b) preset count is greater than the accumulated count.
c) counter input rung is true.
d) counter input rung is false.

4.___A___

5. Which of the following is *not* usually associated with a PLC counter instruction?
a) Address
b) Preset value
c) Time base
d) Accumulated value

5.___C___

6. A PLC up-counter (CTU) counter counts:
a) scan transitions.
b) true-to-false transitions.
c) false-to-true transitions.
d) both b and c.

6.___C___

7. When the up-counter reset is set to true, the following happens: 7. _____

a) the preset value is set to 0.

b) the preset value increments.

c) the accumulated value is set to 0.

d) the accumulated value is set to maximum.

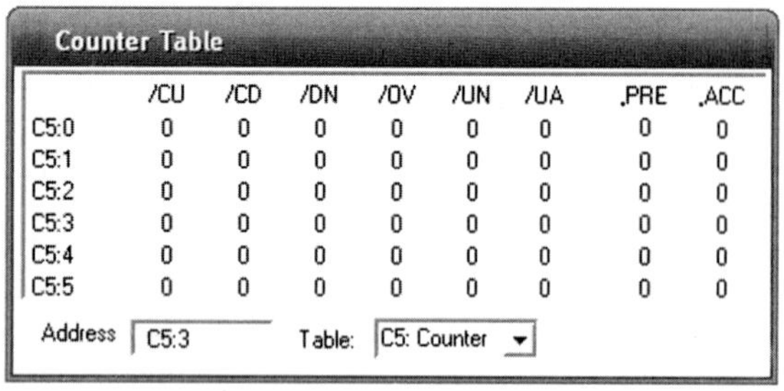

Counter Table

	/CU	/CD	/DN	/OV	/UN	/UA	.PRE	.ACC
C5:0	0	0	0	0	0	0	0	0
C5:1	0	0	0	0	0	0	0	0
C5:2	0	0	0	0	0	0	0	0
C5:3	0	0	0	0	0	0	0	0
C5:4	0	0	0	0	0	0	0	0
C5:5	0	0	0	0	0	0	0	0

Address C5:3   Table: C5: Counter ▼

Figure 8-1 Counter table for question 8.

8. For the counter table shown in Figure 8-1, word level addressing 8. _____
is used for:

a) CU, CD, DN, OV, UN, and UA.   c) OV and UN.

b) CU, CD, and DN.   d) PRE and ACC.

9. In an up-counter, when the accumulated count exceeds the preset count 9. _____
without a reset, the accumulated count will:

a) set itself to zero.   c) continue incrementing.

b) start decrementing.   d) hold the accumulated value.

10. When the accumulated count exceeds the preset count, the: 10. _____

a) accumulated value is set to zero.   c) reset changes state.

b) preset is set to zero.   d) counter done bit is true.

11. The counter RES instruction: 11. _____

a) is used to reset the counter.

b) is given the same reference address as the counter instruction.

c) decrements the count when actuated.

d) both a and b.

12. For the PLC counter to reset, the counter reset rung must: 12. _A_

a) be true.

b) be false.

c) be either true or false, depending on the manufacturer.

d) undergo a true-to-false transition.

13. Normally counters are retentive. This means that if your accumulated 13. _C_
count is up to 300 and power to your system is lost, when power is restored,
the accumulated count will be:

a) 000.   c) 300.

b) 250.   d) 999.

14. A one-shot, or transitional, contact: 14. _D_

a) operates the same as an NO contact instruction.

b) operates the same as an NC contact instruction.

c) operates the same as a timed closed contact.

d) closes for only one program scan when actuated.

15. A PLC down-counter (CTD) counter counts: 15. _C_

a) scan transitions.   c) false-to-true transitions.

b) true-to-false transitions.   d) both b and c.

16. The accumulated count of a CTD counter: 16. _C_

a) increments with each true-to-false transition.

b) decrements with each true-to-false transition.

c) decrements with each false-to-true transition.

d) increments with each false-to-true transition.

17. The accumulated count of a CTU counter: 17. _D_

a) increments with each true-to-false transition.

b) decrements with each true-to-false transition.

c) decrements with each false-to-true transition.

d) increments with each false-to-true transition.

18. A counter is to be programmed to keep track of the number of parts coming off a production line. If you wanted to subtract the number of rejected parts so your counter would count only the good parts, you would program:

a) two up-counters.

b) two down-counters.

c) an up/down-counter.

d) a counter with a transitional contact input.

18. _C_

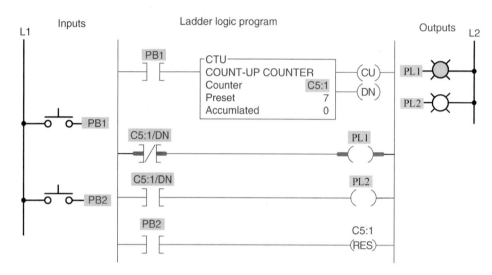

Figure 8-2 Counter program for question 19.

19-1. For the counter program of Figure 8-2, output PL2 will be energized:

a) until the accumulated value equals the preset value.

b) when the accumulated value equals the preset value.

c) only when the accumulated value exceeds 10.

d) only when the accumulated value is zero.

19-1. _b_

19-2. Output PL1 will be energized:

a) until the accumulated value equals the preset value.

b) when the accumulated value equals the preset value.

c) only when the accumulated value is less than 10.

d) only when the accumulated value is 99.

19-2. _A_

19-3. The field device that will cause the counter to increment is: 19-3. _A_

a) input PB1.   c) output PL1.

b) input PB2.   d) output PL2.

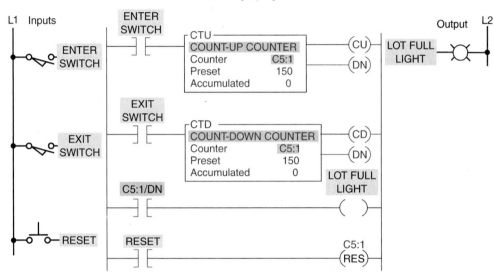

Figure 8-3 Parking garage counter program for question 20.

20-1. For the parking garage counter program of Figure 8-3, the output Lot   20-1. _B_
Full Light will be energized when the accumulated count is:

a) 0.   c) 100.

b) 150.   d) 125.

20-2. Which instruction will cause the counter to decrement?   20-2. _B_

a) Enter switch input.   c) Reset input.

b) Exit switch input.   d) Lot Full Light output.

20-3. Which instruction, when true, will preset the counter to a count   20-3. _C_
of zero?

a) Enter switch input.   c) Reset input.

b) Exit switch input.   d) Lot Full Light output.

20-4. Suppose the accumulated count is 60 before the Enter switch input   20-4. _D_
is actuated 15 times and the Exit switch input is actuated 5 times. After this
operational sequence, the accumulated count would be:
a) 80.   c) 75.
b) 65.   d) 70.

20-5. During normal operation of the program, the accumulated value   20-5. _A_
of CTU would always be:
a) the same as that of CTD.   c) between 50 and 100.
b) 150.   d) between 0 and 100.

20-6. Assume the accumulated count is 100 and the following order   20-6. _C_
of events then occurs: Exit switch input is actuated 20 times, Reset input is
actuated 10 times, and Enter switch input is actuated 5 times. After this
sequence, the accumulated count would be:
a) 100.   c) 5.
b) 115.   d) 0.

21. The counter file for SLC 500 controllers is: 21. _D_
a) C2.   c) C4.
b) C3.   d) C5.

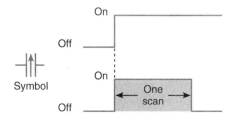

Figure 8-4 Contact instruction for question 22.

22. Figure 8-4 illustrates the operation of a: 22. _C_
a) timer contact instruction.   c) one-shot contact instruction.
b) counter contact instruction.   d) sensor contact instruction.

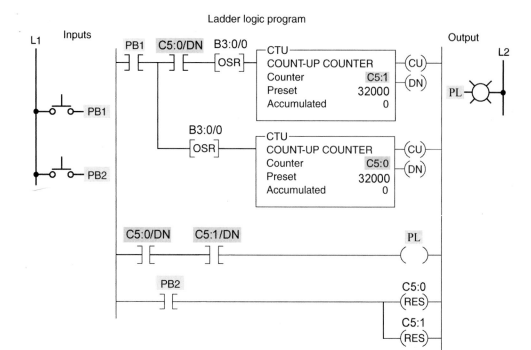

Figure 8-5 Counter program for question 23.

23-1. The counter program of Figure 8-5 is designed to:

a) count up and count down.

b) record the time of an event.

c) count beyond the maximum count allowed per counter.

d) count below the maximum count allowed per counter.

23-1. ___C___

23-2. Counter C5:1 starts counting:

a) after the accumulated value of C5:0 reaches 32,000.

b) before the accumulated value of C5:0 reaches 32,000.

c) whenever input PB2 is actuated.

d) either b or c.

23-2. ___A___

23-3. Output PL will be energized when:

a) the accumulated value of C5:0 reaches 32,000.

b) counter C5:0 is reset.

c) the accumulated value of C5:1 reaches 32,000.

d) the accumulated value of C5:0 and C5:1 reaches 32,000.

23-3. ___D___

23-4. When output PL is energized, how many counts have occurred?   23-4. _C_

a) 4,000   c) 64,000

b) 32,000   d) 99,999

23-5. If you wanted output PL to go on after a count of 40,000, you would   23-5. _C_
change the preset count of C5:1 to:

a) 9,999.   c) 8,000.

b) 6,000.   d) 12,000.

23-6. When input PB2 is actuated: 23-6. _D_

a) output PL is switched off.   c) counter C5:1 is reset.

b) counter C5:0 is reset.   d) all of these.

Ladder logic program

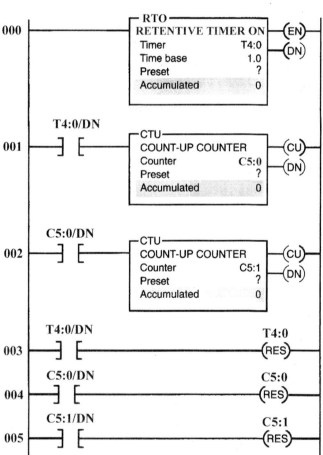

Figure 8-6 Clock
program for question 24.

24-1. The 24-hour clock program of Figure 8-6 uses:

24-1. _C_

a) 3 timers and 3 counters. c) 1 timer and 2 counters.

b) 2 timers and 2 counters. d) 2 timers and 1 counter.

24-2. RTO is preset for a:

24-2. _A_

a) 60-s time period. c) 12-h time period.

b) 2-min time period. d) 24-h time period.

24-3. Counter C5:0 is preset for:

24-3. _C_

a) 12. c) 60.

b) 24. d) 120.

24-4. Counter C5:1 is preset for a:

24-4. _B_

a) 12. c) 60.

b) 24. d) 120.

24-5. A false-to-true transition of rung 002 increments the clock by:

24-5. ____

a) 1 ms. c) 1 min

b) 1 s. d) 1 h

24-6. Rung 003 undergoes a true-to-false transition once every:

24-6. _A_

a) 60 s. c) hour.

b) 2 min. d) 24 h.

24-7. Assume the accumulated count of counter C5:1 is 14 and that of C5:0 is 10. The correct time of day would be:

24-7. _A_

a) 2:10 p.m. c) 10:14 p.m.

b) 10:14 a.m. d) 2:10 a.m.

TEST 8.2

Place the answers to the following questions in the answer column at the right.

1. Programmed counters can serve the same function as mechanical counters. (True or False)

1. _T_

2. The majority of counters used are classified as _____-counters.

2. _up_

3. Every PLC model offers some form of counter instruction. (True or False)

3. _T_

4. Counters are similar to timers, except that they do not operate on an internal clock. (True or False)

4. _T_

5. Counter instructions can be (a) _____ formatted or (b) _____ formatted.

5a. _Coil_
5b. _Box_

6. The up-counter increments its accumulated value by 1 each time the counter rung makes a(n) _____ transition.

6. _Fot_

7. The output of the counter is energized whenever the accumulated count is less than or equal to the preset count. (True or False)

7. _T_

8. A programmed counter is reset by means of a counter _____ instruction.

8. _____

9. PLC counters are normally nonretentive. (True or False)

9. _F_

10. Some PLC counters operate on the leading edge of the input signal, while others operate on the trailing edge. (True or False)

10. _F_

11. All PLC manufacturers require the reset rung or line to be true to reset the counter. (True or False)

11. _T_

12. A transitional off-to-on contact will allow logic continuity for one scan and then open, even though the triggering signal may stay on. (True or False)

12. _____

13. The acronym CTD stands for a(n) _____ counter instruction.

13. _CD_

14. The acronym CTU stands for a(n) ____ counter instruction.

14._____

15. The ____ instruction is used to set the counter accumulated value to zero.

15._____

16. The transitional contact instruction is also known as a(n) ____ contact instruction.

16._____

17. Transitional contacts are often used for ____ counters and timers.

17. _Reset_

18. A down-counter output instruction will decrement by 1 each time the counted event occurs. (True or False)

18. _T_

19. In normal use, the down-counter is used in conjunction with the up-counter to form an up/down-counter. (True or False)

19. _T_

20. All up-counters count only to their preset values, and additional counts are ignored. (True or False)

20. _T_

21. One way of counting events that exceed the maximum number allowable per counter instruction is by the ____ of two counters.

21. _Cascade_

22. The counter reset (RES) instruction, it is always given the same address as the counter it is to reset. (True or False)

22. _T_

23. The counter enable bit is true whenever the counter instruction is false. (True or False)

23. _F_

24. The counter done bit is true whenever the (a) ____ value is equal to or greater than the (b) ____ value.

24a. _Acc_
24b. _PRS_

25. The counter ____ bit is true whenever the counter counts past its maximum value.

25. _down_

26. The counter ____ values specifies the value that the counter must count to before it changes the state of the done bit.

26. _T_

27. The counter accumulated value is the current count based on the number of times the rung goes from false to true. (True or False)

27._____ T

28. The counter number C5:4 represents counter file 5, counter 4 in that file. (True or False)

28._____ T

29. Encoder pulses can be counted to measure distance. (True or False)

29._____ T

30. A counter instruction is an input instruction. (True or False)

30._____ F

31. A counter's input signal can come from an external device such as a sensor. (True or False)

31._____ T

32. Up and down counters may be programmed together to count up and down. (True or False)

32._____ T

33. Counters can count past their preset values. (True or False)

33._____ T

Programming Assignments

This section presents several common counter program applications. The instructions used are intended to be generic in nature and, as such, will require some conversion for the particular PLC model you are using. The use of a prewired PLC input/output control panel is recommended to simulate the operation of these circuits.

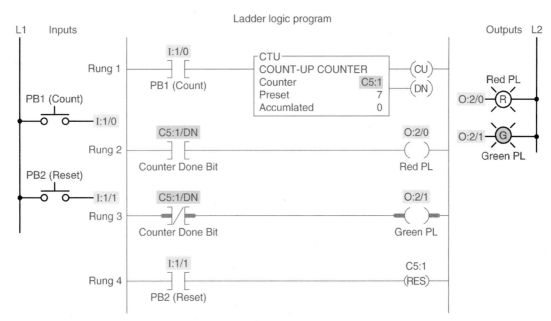

Figure 8-7 Up-counter program for assignment 1.

1) The operation of the up-counter program shown in Figure 8-7 is described in the text. Prepare an I/O connection diagram and ladder logic program for the process. Use addresses that apply to your PLC. Enter the program into the PLC, and prove its operation.

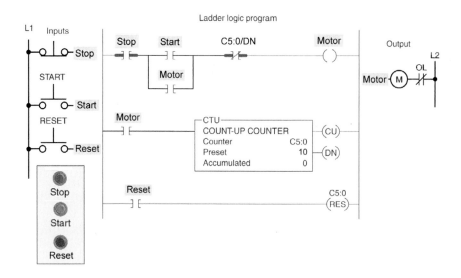

Figure 8-8 Motor counter program for assignment 2.

2) The counter program of Figure 8-8 is used to stop a motor from running after 10 operations. The operation of the program is described in the text. Prepare an I/O connection diagram and ladder logic program for this application. Use addresses that apply to your PLC. Enter the program into the PLC, and prove its operation.

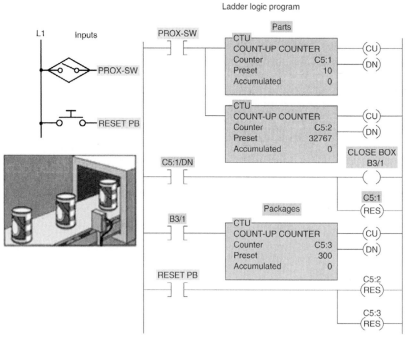

Figure 8-9 Can counting program for assignment 3.

3) The can counting program of Figure 8-9 is described in the text. Prepare an I/O connection diagram and ladder logic program for this process. Use addresses that apply to your PLC. Enter the program into the PLC, and prove its operation.

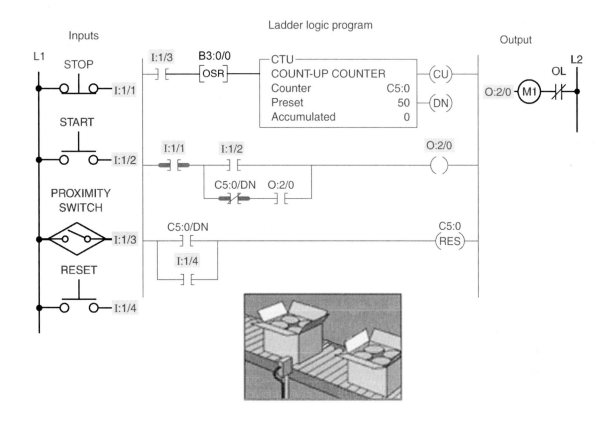

Figure 8-10 Case counting program for assignment 4.

4) The case counting program of Figure 8-10 is described in the text. Prepare an I/O connection diagram and ladder logic program for this process. Use addresses that apply to your PLC. Enter the program into the PLC, and prove its operation.

Ladder logic program

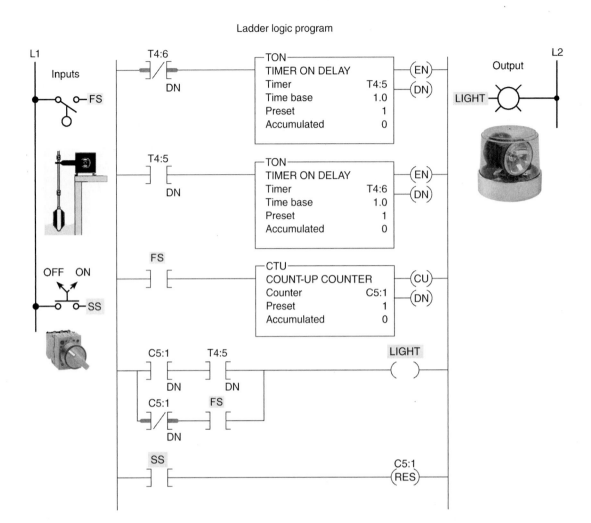

Figure 8-11 Alarm monitor program for assignment 5.

5) The alarm monitor program of Figure 8-11 is described in the text. Prepare an I/O connection diagram and ladder logic program for this process. Use addresses that apply to your PLC. Enter the program into the PLC, and prove its operation.

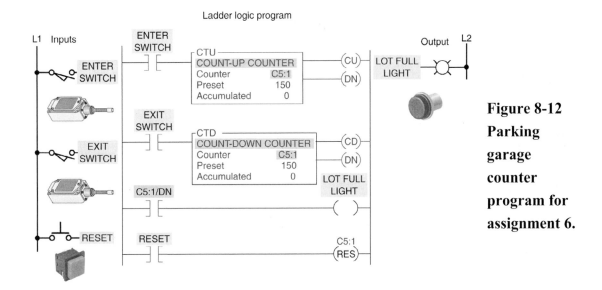

Figure 8-12 Parking garage counter program for assignment 6.

6) The parking garage counter program of Figure 8-12 is described in the text. Prepare an I/O connection diagram and ladder logic program for this process. Use addresses that apply to your PLC. Enter the program into the PLC, and prove its operation.

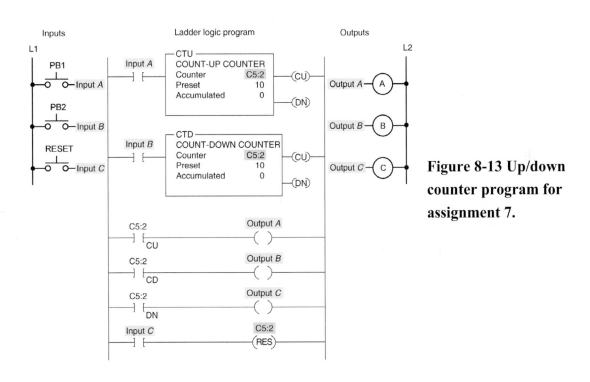

Figure 8-13 Up/down counter program for assignment 7.

7) The operation of the up/down-counter program shown in Figure 8-13 is described in the text. Prepare an I/O connection diagram and ladder logic program for the process. Use addresses that apply to your PLC. Enter the program into the PLC, and prove its operation.

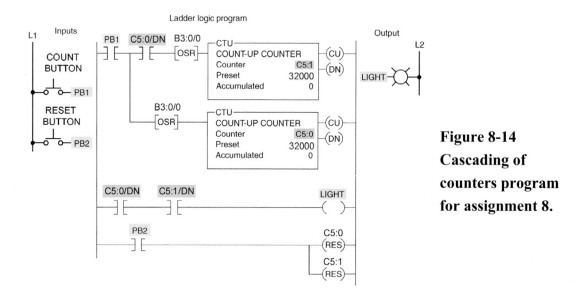

Figure 8-14 Cascading of counters program for assignment 8.

8) The cascading of counters program shown in Figure 8-14 is described in the text. Prepare an I/O connection diagram and ladder logic program for the process. Use addresses that apply to your PLC. Enter the program into the PLC, and prove its operation.

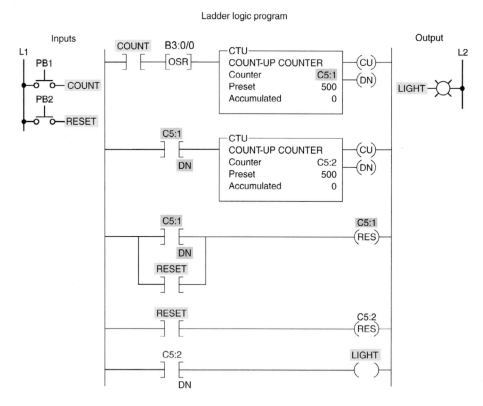

Figure 8-15 Cascading of counters for extremely large counts program for assignment 9.

9) The cascading of counters for extremely large counts program shown in Figure 8-15 is described in the text. Prepare an I/O connection diagram and ladder logic program for the process. Use addresses that apply to your PLC. Enter the program into the PLC, and prove its operation.

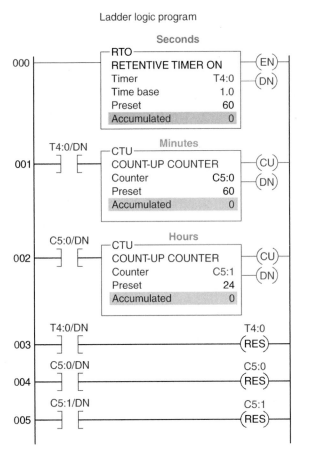

Figure 8-16 24-hour clock program for assignment 10.

10) The 24-hour clock program shown in Figure 8-16 is described in the text. Prepare an I/O connection diagram and ladder logic program for the application. Use addresses that apply to your PLC. Enter the program into the PLC, and prove its operation.

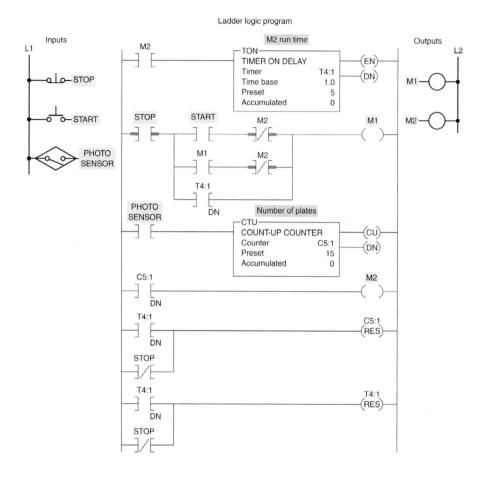

Figure 8-17 Automatic stacking program for assignment 11.

11) The automatic stacking program shown in Figure 8-17 is described in the text. Prepare an I/O connection diagram and ladder logic program for the application. Use addresses that apply to your PLC. Enter the program into the PLC, and prove its operation.

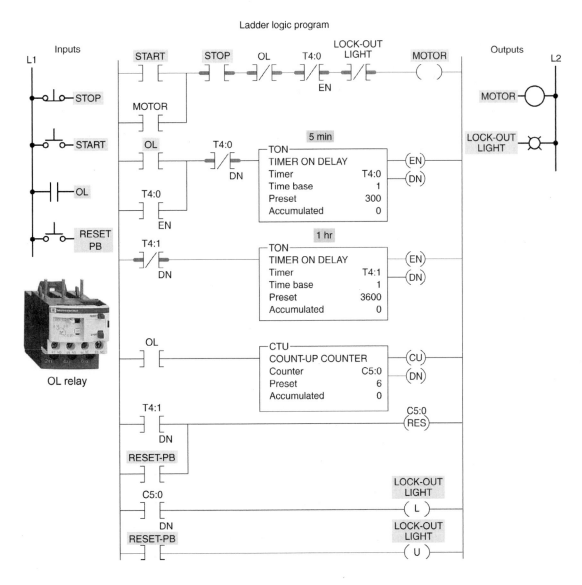

Figure 8-18 Motor lock-out program for assignment 12.

12) The motor lock-out program shown in Figure 8-18 is described in the text. Prepare an I/O connection diagram and ladder logic program for the application. Use addresses that apply to your PLC. Enter the program into the PLC, and prove its operation.

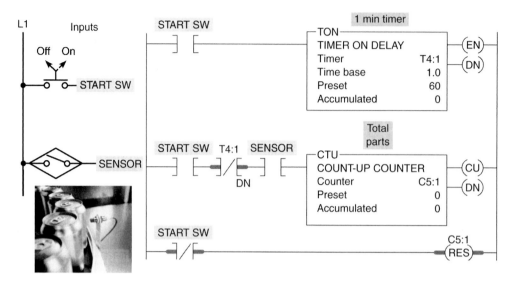

Figure 8-19 Product flow rate program for assignment 13.

13) The product flow rate program shown in Figure 8-19 is described in the text. Prepare an I/O connection diagram and ladder logic program for the application. Use addresses that apply to your PLC. Enter the program into the PLC, and prove its operation.

14) Write a PLC program and prepare a typical I/O connection diagram and ladder logic program for the following counter specifications. Use addresses that apply to your PLC. Enter the program into the PLC, and prove its operation.

- Counts the number of times a pushbutton is closed.
- Decrements the accumulated value of the counter each time a second pushbutton is closed.
- Turns on a light any time the accumulated value of the counter is less than 30.
- Turns on a second light when the accumulated value of the counter is equal to or greater than 40.
- Resets the counter to 0 when a selector switch is closed.

15) Design a PLC program and prepare an I/O connection diagram and ladder logic program that will correctly execute the conveyor motor control process illustrated in Figure 8-20. Enter the program into the PLC, and prove its operation. The operational sequence can be summarized as follows:

- The start button is pressed to start the conveyor motor.
- Cases move past the proximity switch and increment the counter's accumulated value.
- After a count of 50, the conveyor motor stops automatically and the counter's accumulated value is reset to zero.
- The conveyor motor can be stopped and started manually at any time without loss of the accumulated count.
- The accumulated count of the counter can be reset manually at any time by means of the counter reset button.

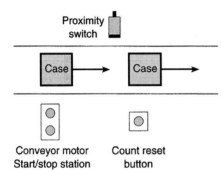

Figure 8-20 Control process for assignment 15.

16) Design a PLC program that will latch on an output, PL1, after an input, PB1, has cycled on 20 times. When the count of 20 is reached, the counter will reset itself automatically. PB2 will unlatch PL1. Enter the program into the PLC, and prove its operation.

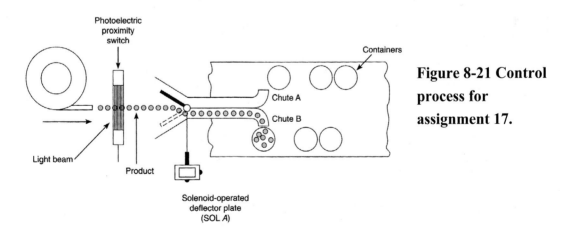

Figure 8-21 Control process for assignment 17.

17) Design a PLC program and prepare an I/O connection diagram and ladder logic program that will correctly execute the packaging process illustrated in Figure 8-21. Enter the program into the PLC, and prove its operation. The operational sequence can be summarized as follows:

- The purpose of this process is to deposit 50 pieces of the product in each container.
- The process is set in operation by pressing a start pushbutton.
- As the product passes through the light beam, it is detected by the photoelectric proximity switch and counted by the PLC counter.
- When the count reaches 50, the solenoid-operated deflector plate (SOL A) energizes to channel the product from chute A to chute B.
- The counter is reset automatically for the next count of 50.
- When the second count of 50 is reached, the solenoid-operated deflector plate de-energizes to channel the product back into chute A, and so on.
- Provisions must be made for stopping the process at any time and manually resetting the accumulated value of the counter to any number.

18) Write a program to operate a light according to the following sequence. Enter the program into the PLC, and prove its operation.

- A momentary pushbutton is pressed to start the sequence.
- The light is switched on and remains on for 5 s.
- The light is then switched off and remains off for 5 s.
- A counter is incremented by 1 after this sequence.
- The sequence then repeats for a total of 5 counts.
- After the fifth count, the sequence will stop and the counter will be reset to zero.

19) Write a program designed to alternate the use of two input pumps so that they both get the same amount of usage over their lifetime. Enter the program into the PLC, and prove its operation. The control process can be summarized as follows:

- A start/stop pushbutton station is provided for control of the two input pump motors P1 and P2.
- The start/stop pushbutton station is operated to control pump P1.
- When the tank is full, drain pump motor P3 is started automatically and runs until the low level sensor is actuated.
- After 5 fillings of tank by pump P1, control automatically shifts to pump P2.
- Operation of the start/stop pushbutton now controls pump P2.
- After 5 fillings of the tank by pump P2, the sequence is repeated.

TEST 9.1

Choose the letter that best completes the statement.

1. Which of the following PLC instructions would *not* be classified as an override instruction?

a) Master control reset   c) Output energize

b) Jump-to-subroutine   d) Jump-to-label

1._____

2. The MCR instruction establishes a zone in the user program in which all nonretentive outputs can be:

a) turned on simultaneously.   c) turned on in a defined sequence.

b) turned off simultaneously.   d) turned off in a defined sequence.

2._____

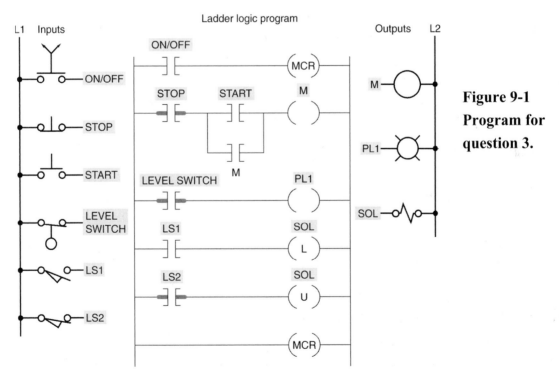

Figure 9-1
Program for
question 3.

3-1. In the program of Figure 9-1, when the MCR instruction is false, output(s) _____ will always be deenergized.

a) M, PL1, and SOL   c) PL1

b) M and PL1   d) SOL

3-1._____

3-2. Assume that the MCR instruction makes a false-to-true transition.   3-2._____

As a result:

a) all outputs will be controlled by the respective input conditions.

b) all nonretentive outputs will deenergize.

c) all retentive outputs will deenergize.

d) all nonretentive outputs will energize.

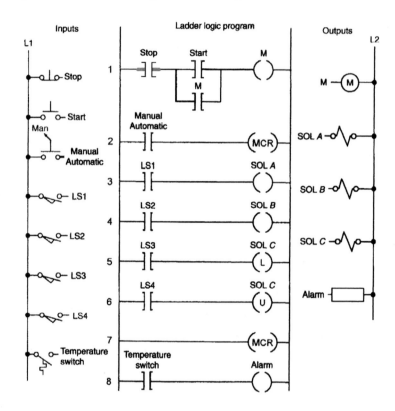

Figure 9-2
Program for
question 4.

4-1. In the program in Figure 9-2, assume output SOL C is energized   4-1._____
at the time the MCR instruction makes a true-to-false transition. As a result,
output SOL C will:

a) deenergize.

b) remain energized.

c) remain energized but still be controlled by inputs LS3 and LS4.

d) deenergize but still be controlled by inputs LS3 and LS4.

4-2. The fenced zone controlled by the MCR instruction is (are):   4-2._____

a) rungs No. 1 through No. 8.

b) rungs No. 3 through No. 6.

c) rungs No. 2 and No. 6.

d) rung No. 2.

4-3. Which of the following is the conditional instruction that controls the MCR zone?   4-3._____

a) Stop

b) Start

c) Manual Automatic

d) Temperature Switch

4-4. The latch and unlatch instructions would be classified as: 4-4._____

a) retentive outputs.

b) nonretentive outputs.

c) conditional instructions.

d) unconditional instructions.

4-5. Assume the alarm output is activated. This would require the: 4-5._____

a) temperature switch input to be true.

b) Manual Automatic input to be true.

c) LS4 input to be true.

d) both a and b.

5. The main advantage to the jump-to-label instruction is that: 5._____

a) any number of rungs may be programmed between the jump and label rungs.

b) it allows you to use one set of condition instructions to control multiple outputs.

c) it allows you to use one set of condition instructions to control multiple inputs.

d) it has the ability to reduce the processor scan time.

6. The label (LBL) instruction is: 6._____

a) always logically true.

b) has the same address as the jump instruction with which it is used.

c) is used to identify the ladder rung that is the target destination of the JMP instruction.

d) all of these.

7. Which of the following instructions would most likely be programmed outside the jumped area of a program?

7._____

a) Latch and unlatch instructions

b) Timer and counter instructions

c) Immediate inputs and outputs

d) Forced inputs and outputs

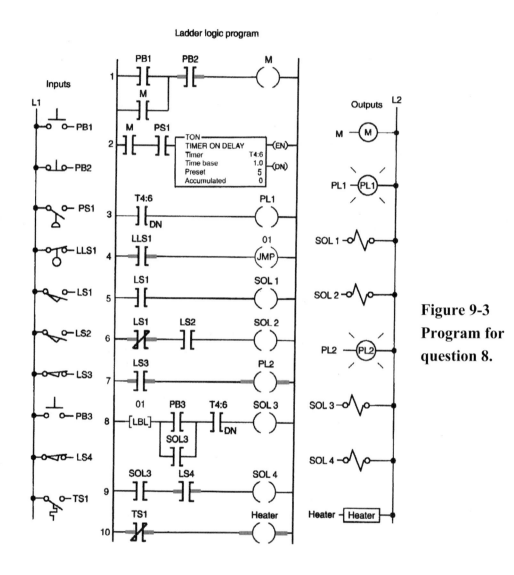

Ladder logic program

Figure 9-3 Program for question 8.

8-1. In the program of Figure 9-3, which output is not affected by the jump instruction?

8-1._____

a) M

c) SOL 2

b) SOL 1

d) PL2

168

8-2. Rungs 5, 6, and 7 are not scanned by the processor when rung
_____ has logic continuity.

8-2._____

a) 1   c) 3
b) 2   d) 4

8-3. When the jump-to-label instruction is executed, the outputs of the
jumped rungs:

8-3._____

a) are all energized.   c) are all immediately updated.
b) are all deenergized.   d) remain in their last state.

9. The jump-to-subroutine instruction can save a great deal of duplicate
programming in cases:

9._____

a) that require the programming of several timers.
b) that require the programming of several counters.
c) where a machine has a portion of its cycle that must be repeated several
times during one machine cycle.
d) all of these.

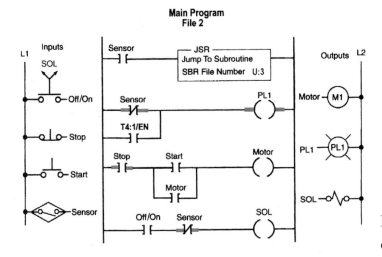

Figure 9-4 Program for question 10.

169

10-1. In the program of Figure 9-4, when the examine on sensor instruction is true, the processor:

10-1.____

a) turns on all outputs in the subroutine area.

b) turns off all outputs in the subroutine area.

c) stops executing the subroutine.

d) starts executing the subroutine.

10-2. When the processor scan reaches the RET instruction, it will return the processor to the:

10-2.____

a) start of the program.   c) rung above the JSR instruction.

b) end of the program.   d) rung below the JSR instruction.

11. The immediate input and output instructions provide a way of:

11._____

a) ending the program immediately.

b) restarting the program immediately.

c) temporarily interrupting the program scan to allow selected bits in the data table to be updated.

d) temporarily interrupting the program scan to reset all bits in the data table to zero.

12. Immediate instructions should be used only when:

12._____

a) a program must be halted immediately.

b) a program must be restarted immediately.

c) the updating of an input or output is critical to your operation.

d) the resetting of all bits in the data table is critical to your operation.

13. Immediate instructions are most useful when programmed:

13._____

a) immediately after the I/O scan has occurred.

b) immediately prior to the I/O scan.

c) at the middle or toward the end of the program.

d) near the beginning of the program.

14. The use of immediate instructions:

14._____

a) increases the total scan time of the program.

b) decreases the total scan time of the program.

c) increases the number of rungs that can be programmed.

d) decreases the number of rungs that can be programmed.

15. The forcing function of a PLC allows the user to turn an external input   15._____
or output on or off:

a) according to the forced program.

b) according to the main program.

c) from the keyboard regardless of its actual status.

d) all of these.

16. Forcing functions are often used: 16._____

a) to continue a machine process until a faulty field device can be repaired.

b) for testing purposes during an initial start-up.

c) for troubleshooting purposes.

d) all of these.

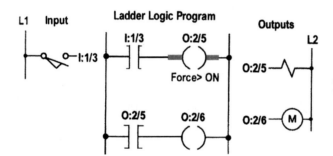

Figure 9-5 Program for question 17.

17-1. In the program of Figure 9-5, the actual status of input I:1/3 is _____   17-1.____
but the forced status is _____.

a) false, true   c) false, false

b) true, false   d) true, true

17-2. The output of O:2/5 would be _____ and the output of O:2/6   17-2.____
would be _____.

a) false, true   c) false, false

b) true, false   d) true, true

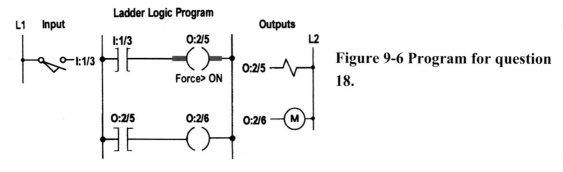

Ladder Logic Program

Figure 9-6 Program for question 18.

18-1. In the program of Figure 9-6, the actual status of output address O:2/5 is ____, but the forced status is ____.

18-1.____

a) false, true
b) true, false
c) false, false
d) true, true

18-2. The status of examine on instruction O:2/5 would be ____ and the status of output O:2/6 would be ____.

18-2.____

a) false, true
b) true, false
c) false, false
d) true, true

19. Forcing functions should not be used:

19._____

a) with retentive outputs.
b) with nonretentive outputs.
c) with immediate I/O instructions.
d) without consideration for any potential unsafe effects.

20. PLC emergency stop circuits should be:

20._____

a) hardwired outside the controller program.
b) programmed as part of the master control reset instruction.
c) programmed as part of the zone control last state instruction.
d) programmed as an immediate input instruction.

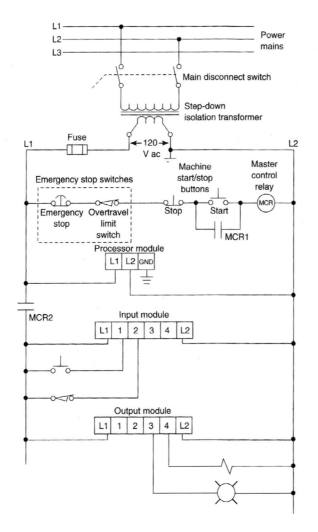

Figure 9-7 Program for question 21.

21-1. In the diagram of Figure 9-7, power to the processor module is controlled by the:

21-1.____

a) master control relay.

c) main disconnect switch.

b) start/stop buttons.

d) all of the above.

21-2. Power to the input and output module is controlled by the:

21-2.____

a) master control relay.

c) main disconnect switch.

b) start/stop buttons.

d) all of the above.

21-3. The transformer is used to:

21-3.____

a) isolate the controller from the main power lines.

b) step up the main power line voltage.

c) provide the low-voltage operating voltage for the controller.

d) all of the above

21-4. Assume the processor comes equipped with a normally closed fault   21-4._____
relay contact output designed to open when a processor malfunction is
detected. This contact would be:
a) programmed as part of the master control reset instruction.
b) programmed as part of the zone control last state instruction.
c) hardwired in series with the emergency stop button.
d) hardwired in parallel with the emergency stop button.

21-5. When replacing modules or working on equipment controlled by the   21-5._____
PLC installation, the safest way to proceed is to:
a) deenergize the MCR coil.
b) block open the emergency stop switch.
c) remove the fuse from the circuit.
d) pull and lock the disconnect switch.

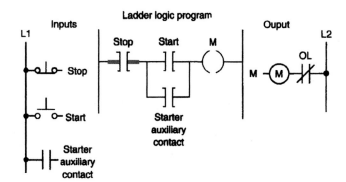

Figure 9-8 Program for question 22.

22-1. In the program in Figure 9-8, the use of the starter auxiliary contact   22-1._____
instead of a programmed contact:
a) is more costly.
b) is safer.
c) provides positive feedback about the exact status of the motor.
d) all of these.

22-2 Assume that the stop button was changed to a normally open contact   22-2._____
type. As a result, the program could be made to operate as before by
changing the:
a) stop instruction to examine if open.
b) start instruction to examine if open.
c) starter auxiliary contact instruction to examine if open.
d) both a and c.

22-3. Assume that the NC stop button is replaced with an NO stop button   22-3.____
and that the program is changed so it operates as before. Should the field
wire connected to one end of the stop button break off:

a) the motor would stop automatically.

b) pressing the stop button would stop the motor.

c) pressing both the start and stop buttons would stop the motor.

d) none of these.

23. A jump instruction is similar to a(n):   23._____

a) MCR command.   c) skip command.

b) ZCL command.   d) JSR command.

24. A JSR instruction: 24._____

a) tells the processor to jump from the main program to a subroutine area or file.

b) tells the processor to execute the fault routine.

c) latches outputs when energized.

d) latches outputs when deenergized.

25. The MCR instruction: 25._____

a) is an output instruction.

b) is used in pairs.

c) is used to disable or enable a zone within a ladder program.

d) all of these.

26. The ____ is the target for the jump instruction.   26._____

a) LBL   c) IOT

b) TND   d) RET

27. The ____ instruction will return the scan to your main program at the   27._____
completion of the subroutine.

a) LBL   c) TND

b) IIN   d) RET

28. The _____ instruction stops the processor from scanning the rest
of the program.
a) LBL   c) TND
b) IOT   d) STI

28._____

29. The MCR instruction can be used to control:
a) entire sections of a program.
b) entire rungs of a program.
c) selected elements within a rung of a program.
d) both a and b.

29._____

30. The JMP and LABEL instructions allow a processor to:
a) return to the beginning of a program.
b) take a shortened route to the end of the program.
c) skip sections of the program, reducing scan time.
d) create a fault condition.

30._____

31. The JSR instruction requires that a(n):
a) separate file be created.   c) separate power supply be used.
b) separate processor be used.   d) all of these.

31._____

TEST 9.2

Place the answers to the following questions in the answer column at the right.

1. Master control reset (MCR) and jump (JMP) are often referred
to as ____ instructions.

1._____

2. The MCR instruction can only be programmed to control an entire
circuit. (True or False)

2._____

3. When the MCR instruction is ____, all rung outputs below the
MCR will be controlled by their respective input conditions.

3._____

4. If the MCR output is turned off or deenergized, all nonretentive
rungs below the MCR will be ____.

4._____

5. ____ instructions should not normally be placed within an MCR zone
because they will remain in their last active state when the instruction
goes false.

5._____

6. When programming an MCR instruction to control a fenced zone,
an MCR rung with no conditional inputs is placed at the beginning of the
zone and an MCR rung with conditional inputs is placed at the end of the
zone. (True or False)

6._____

7. The master control instruction is used as a substitute for a hardwired
emergency stop switch. (True or False)

7._____

8. The jump (JMP) instruction is used to jump over certain program
instructions if certain conditions exist. (True or False)

8._____

9. The advantage of the JMP instruction is the ability to reduce
the processor ____.

9._____

10. In a jump-to-label program, the ____ instruction is used to identify
the ladder rung that is the target destination of the jump instruction.

10._____

11. The label address must match that of the jump instruction with which it is used. (True or False)

11._____

12. The JMP instruction does not contribute to logic continuity and, for all practical purposes, is always logically true. (True or False)

12._____

13. The jump-to-subroutine instruction is used where a machine has a portion of its cycle that must be ____ several times during one machine cycle.

13._____

14. When the program scan reaches an immediate I/O instruction, the scan is interrupted and the bits of the addressed word are ____.

14._____

15. The immediate I/O instruction is used with ____ I/O devices that require updating in advance of the I/O scan.

15._____

16. The immediate I/O instruction is most useful if the instruction associated with the device is at the beginning of the program. (True or False)

16._____

17. The use of the immediate I/O instruction increases the total ____ of the program.

17._____

18. The forcing capability of a PLC allows the user to turn an external I/O on or off regardless of the ____ of the device.

18._____

19. Random forcing of given inputs or outputs can cause equipment damage. (True or False)

19._____

20. Emergency stop circuits should be ____ outside of the controller program so that, in the event of total controller failure, independent and rapid shutdown means are available.

20._____

21. A main ____ is installed on the incoming power lines as a means of removing power from the entire PLC system.

21._____

22. Power to the PLC input and output devices is usually controlled by means of a hardwired ____ circuit.

22._____

23. The master control relay instruction can be used as a substitute for a disconnect switch. (True or False)

23._____

24. The use of a motor starter seal-in contact in place of a programmed contact provides _____ feedback about the exact status of the motor.

24._____

25. The safest way to wire a stop button to a PLC system is to use a(n) (a) _____ contact programmed to examine for a(n) (b) _____ condition.

25a._____
25b._____

26. The label instruction has a logical true condition. (True or False)

26._____

27. Jumping to a subroutine does not cause any rungs of the main program to be skipped over. (True or False)

27._____

28. The jump instruction allows a section of a program to be jumped when a production fault occurs. (True or False)

28._____

29. It is not possible to jump backward in the program. (True or False)

29._____

30. Nesting subroutines allow you to direct program flow from the (a) _____ program to a subroutine and then to another (b) _____.

30a._____
30b._____

31. Nested subroutines make complex programming easier. (True or False)

31._____

32. Forcing outputs affects only the addressed output terminal. (True or False)

32._____

33. Programming the selectable timed interrupt is done when a section of program needs to be executed on a(n) _____ basis rather than on an event basis.

33._____

34. The fault routine allows for an orderly shutdown in case of a fault. (True or False)

34._____

35. The temporary end instruction, when true, _____ the program scan.

35._____

36. A latch instruction will automatically unlatch if it is contained within a deenergized MCR zone. (True or False)

36._____

37. The MCR instruction is not a replacement for a hardwired master control relay that provides emergency stop. (True or False)

37._____

38. When programming MCR instructions, the first rung has a conditional MCR output instruction and the last rung is a(n) ____ MCR rung.

38._____

39. When an MCR zone goes false, off-delay timers within the zone will automatically activate and begin their off-delay cycle. (True or False)

39._____

40. Each JMP instruction must have a LBL instruction. (True or False)

40._____

41. You must deenergize the JMP instruction to activate it. (True or False)

41._____

42. You must energize the MCR instruction to activate it. (True or False)

42._____

Programming Assignments

This section requires you to simulate several program control applications. The instructions used are intended to be generic in nature and, as such, will require some conversion for the particular PLC model you are using. The use of a prewired PLC input/output control panel is recommended to simulate the operation of these circuits.

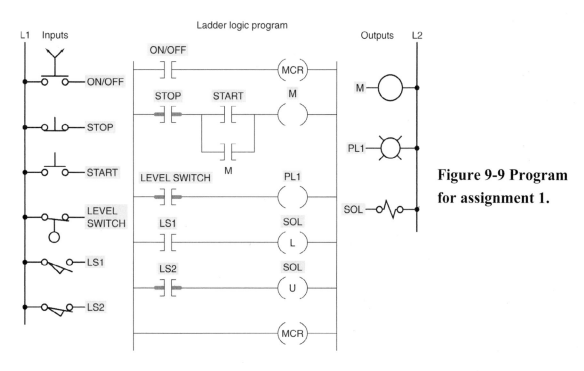

Figure 9-9 Program for assignment 1.

1) Construct a simulated program for the MCR program of Figure 9-9 using any available addresses, switches, and lights on your PLC control panel. After constructing your program on paper, enter the program into the PLC. Demonstrate that when the MCR instruction is deenergized, all nonretentive outputs deenergize and all retentive outputs remain in their last state.

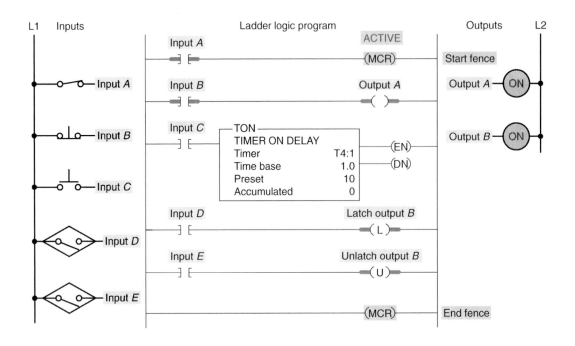

Figure 9-10 Program for assignment 2.

2) Construct a simulated program for the MCR fenced zone program of Figure 9-10. After constructing your program on paper, enter it into the PLC. Demonstrate how the rungs between the two MCR instructions are controlled.

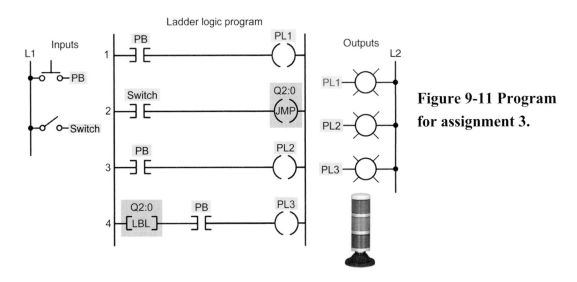

Figure 9-11 Program for assignment 3.

3) Construct a simulated program for the jump-to-label program of Figure 9-11. After constructing your program on paper, enter it into the PLC. Demonstrate how the jump-to-label operation is executed.

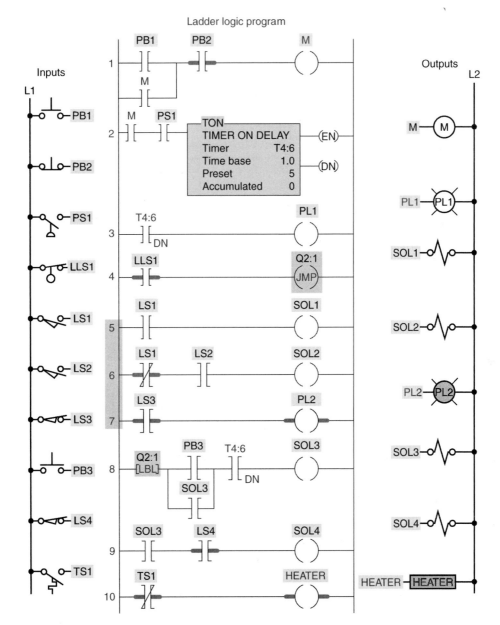

Figure 9-12 Program for assignment 4.

4) The jump-to-label program of Figure 9-12 is described in the text. Prepare an I/O connection diagram and ladder logic program that will simulate its operation. Use addresses that apply to your PLC installation. Enter the program into the PLC, and verify its operation.

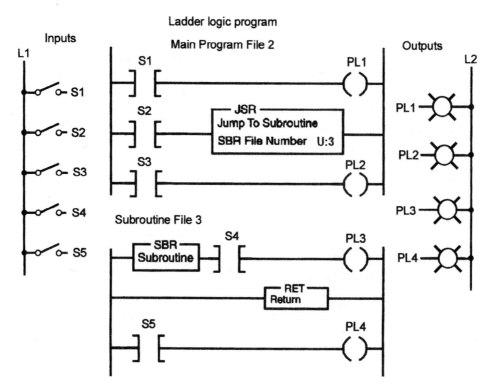

Figure 9-13 Program for assignment 5.

5) Construct a simulated program for the jump-to-subroutine program in Figure 9-13. After constructing your program on paper, enter it into the PLC. Demonstrate how the jump-to-subroutine operation is executed.

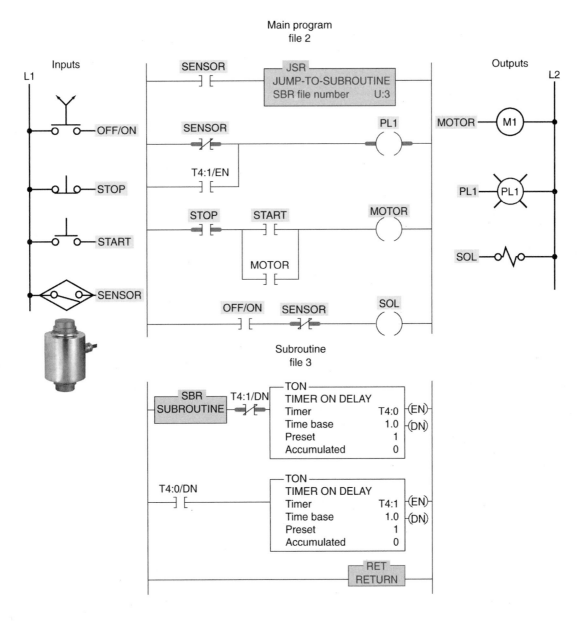

Figure 9-14 Program for assignment 6.

6) The flashing pilot light subroutine program of Figure 9-14 is described in the text. Prepare an I/O connection diagram and ladder logic program that will simulate its operation. Use addresses that apply to your PLC installation. Enter the program into the PLC, and verify its operation.

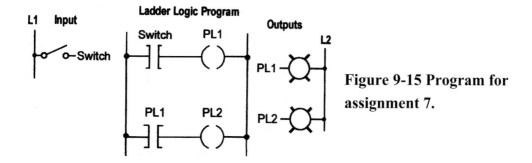

Figure 9-15 Program for assignment 7.

7) Enter the forcing program of Figure 9-15 into the PLC, and demonstrate how each of the following is executed: (a) forcing the switch on; (b) forcing the switch off; (c) forcing PL1 on; and (d) forcing PL1 off.

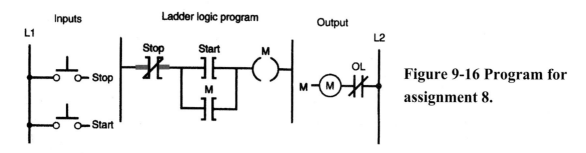

Figure 9-16 Program for assignment 8.

8) Enter the simulated start/stop pushbutton program of Figure 9-16 in the PLC.

- Demonstrate how an open in the stop pushbutton circuit will fail to deenergize the output.
- Replace the normally open stop pushbutton with a normally closed type, and modify the program so that the circuit operates properly. Demonstrate how an open in the stop pushbutton circuit of this program will automatically deenergize the output.

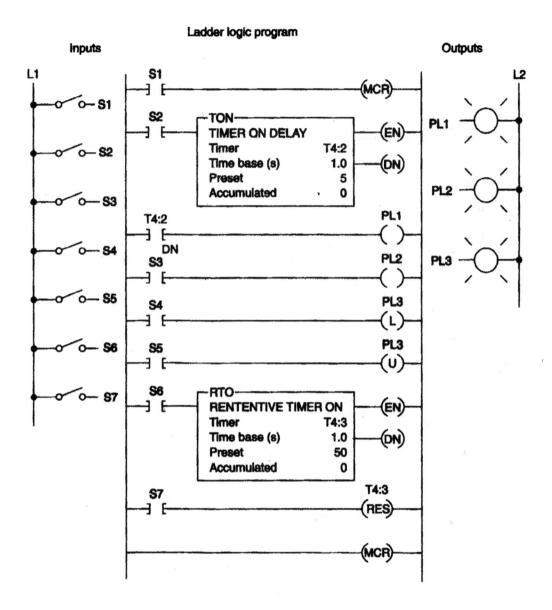

Figure 9-17 Program for assignment 9.

9) Construct a simulated program for the MCR program of Figure 9-17. Enter the program into the PLC. Operate the program according to the following sequence:

a) Close switches 1, 2, 3, 4, and 6, and allow timer T4:2 to time out. What lights are on?

b) Open switch 1. What light is on now? Why did lights 1 and 2 go off?

c) Open switch 4 and close switch 5. Did light 1 go off? Why or why not?

d) What happened to the two timers when you disabled the MCR zone?

e) What happened to the two timers when you reenabled the MCR zone?

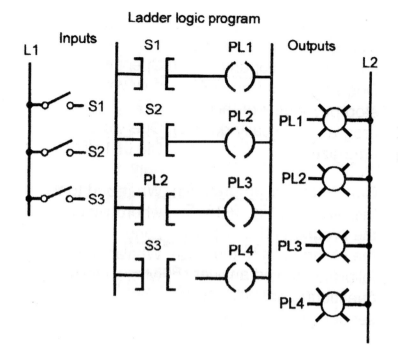

Figure 9-18 Program for assignment 10.

10) Construct a simulated program for the ladder logic program of Figure 9-18. Enter the program into the PLC. Operate the program to force specified inputs and outputs according to the following sequence:

a) Turn switches 1 and 2 off and switch 3 on.

b) Force off switch 3. Use the data monitor to observe the status of the corresponding bit for switch 3 in the input image table file. Close switch 3 and observe the status of the bit. How does forcing inputs manipulate the input image table file bits? Disable the force, and exit from the data monitor to the ladder logic screen.

c) With all switches turned off, force on pilot light 2. Use the data monitor to observe the status of the corresponding bit for pilot light 2 in the output image table file. How does forcing outputs manipulate the output image table file bits?

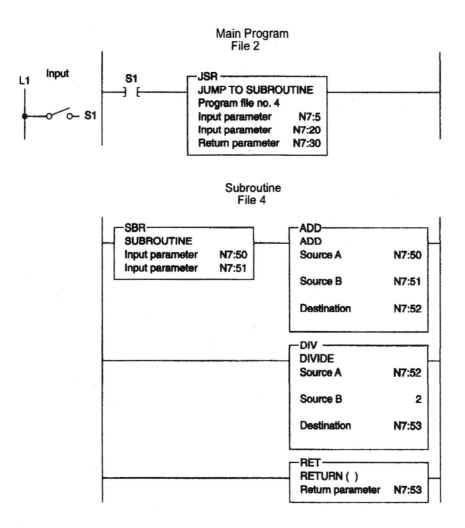

Figure 9-19 Program for assignment 11.

11) Construct the example of the subroutine instruction (SBR) and the return instruction (RET) program shown in Figure 9-19. The purpose of the program is to find the average value of N7:5 and N7:20 and store the result in N7:30. This is accomplished by passing parameters to the subroutine and doing the math in the subroutine, and then returning the answer to the main program through the RET instruction. The operation of the program can be summarized as follows:

- When S1 is closed, the data from the input parameter, N7:5, are copied into the first input parameter in the SBR instruction, N7:50.
- The data from the second input parameter, N7:20, are copied into the second input parameter in the SBR, N7:51.

189

- In the subroutine, N7:50 and N7:51 are then added together, with the result stored in N7:52.
- The value in N7:52 is then divided by 2, which gives the average of N7:50 and N7:51, with the result stored in N7:53.
- The RET instruction then returns the average value through the return parameter, N7:53, to N7:30 in the JSR instruction in the first rung in the main program.

Prove the operation by using the data monitor to insert values for N7:5 and N7:20 and verifying that the average value is contained in N7:30.

12) Design a program that uses the temporary end (TND) instruction described in the text. Demonstrate how this instruction can be used to progressively debug the program.

CHAPTER **10**   **Data Manipulation Instructions**

TEST 10.1

Choose the letter that best completes the statement.

1. Data manipulation instructions enable the PLC to   1._____
a) move data from one memory area to another.
b) compare data.
c) take on some of the qualities of a computer.
d) all of these

2. Depending on the manufacturer, which of the following might be   2._____
considered the same as a word?
a) Register   c) Table
b) File   d) All of these

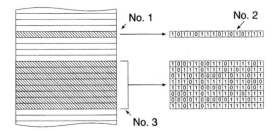

Figure 10-1 Memory map for question 3.

3. According to the memory map of Figure 10-1,   3._____
a) No. 1 is a word, No. 2 is a register, No. 3 is a file.
b) No. 1 is a register, No. 2 is a bit, No. 3 is a file.
c) No. 1 is a file, No. 2 is a bit, No. 3 is a table.
d) No. 1 is a bit, No. 2 is a table, No. 3 is a file.

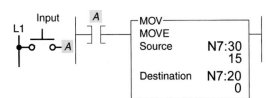

Figure 10-2 Logic rung for question 4.

4-1. The logic rung of Figure 10-2 is an example of a

a) data compare instruction.   c) timer instruction.

b) data move instruction.   d) counter instruction.

4-1._____

4-2. The logic rung is telling the processor to copy the data stored in word

a) N7:30 to word N7:20 when input A is true.

b) N7:20 to word N7:30 when input A is true.

c) N7:30 to word N7:20 when input A is false.

d) N7:20 to word N7:30 when input A is false.

4-2._____

4-3. If input A is closed and then opened

a) 15 will be stored in N7:20 and 0 in N7:30.

b) 0 will be stored in N7:20 and 15 in N7:30.

c) 15 will be stored in N7:20 and 15 in N7:30.

d) 0 will be stored in N7:20 and 0 in N7:30.

4-3._____

5. The masked move (MVM) instruction

a) is an output instruction.

b) moves data through a mask to get to their destination.

c) hides a portion of a binary word before transferring it to its destination.

d) all of these

5._____

6. For the following masked move instruction data, the contents of the destination after the MVM went true would be

6._____

1100 1010 0011 0110 contents of the source

1111 0000 1111 1111 contents of the mask

1010 1101 1010 1010 contents of the destination before MVM went true

a) 1100 1101 001 10110.

b) 1010 1010 1010 1010.

c) 1111 0000 1111 0000.

d) 0000 1111 0000 1111.

7. The bit distribute (BTD) instruction is used to

a) move bits within a word.   c) hide bits of a binary word.

b) move bits between words.   d) both a and b

7._____

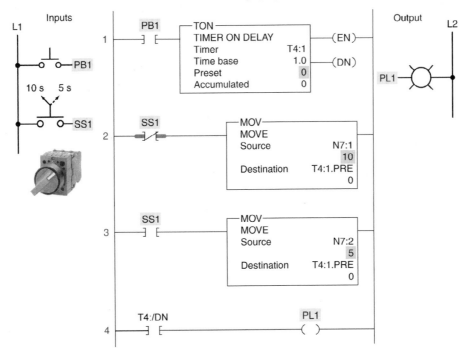

Ladder logic program

Figure 10-3 Timer program for question 8.

8-1. For the timer program of Figure 10-3, the timer starts timing when   8-1._____

a) PB1 is open.   c) SS1 is open.

b) PB1 is closed.   d) SS1 is closed.

8-2. When SS1 is closed, the time-delay period will be   8-2._____

a) 0 s.   c) 10 s.

b) 5 s.   d) 15 s.

8-3. Rung No. 3 will be true   8-3._____

a) 10 s after PB1 remains closed.

b) 5 s after SS1 remains closed.

c) 15 s after both PB1 and SS1 remain closed.

d) any time SS1 is closed.

8-4. Rung No. 2 tells the processor to set the preset time of the timer to   8-4._____

a) 15 when SS1 is open.   c) 10 when SS1 is open.

b) 15 when SS1 is closed.   d) 10 when SS1 is closed.

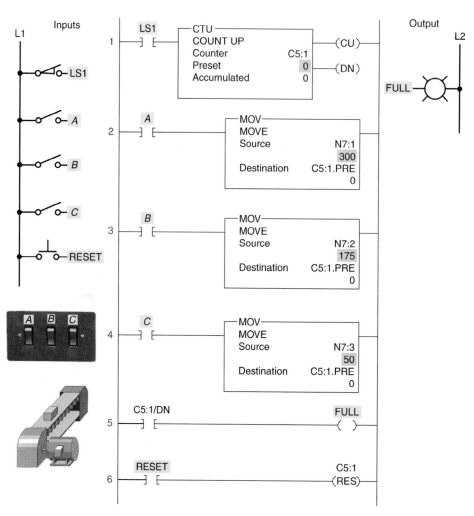

Figure 10-4 Counter program for question 9.

9-1. For the counter program of Figure 10-4, the counter increments by one for each false-to-true transition of

9-1._____

a) rung No. 1.

c) rung No. 3.

b) rung No. 2.

d) rung No. 4.

9-2. A preset count of 50 is selected when input _____ is closed.

9-2._____

a) LS1

c) B

b) A

d) C

9-3. For the light to come on after an accumulated count of 175   9-3._____

a) rung No. 3 must be true.

b) rungs No. 2, No. 3, and No. 4 must be true.

c) rungs No. 1, No. 2, and No. 6 must be true.

d) rungs No. 1, No. 2, No. 4, and No. 6 must be true.

10. A file is a group of   10._____

a) related consecutive words.   c) related instructions.

b) unrelated consecutive words.   d) unrelated instructions.

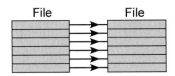

Figure 10-5 Illustration for question 11.

11. The illustration shown in Figure 10-5 illustrates a   11._____

a) word to file move.   c) file to file move.

b) file to word move.   d) file to instruction move.

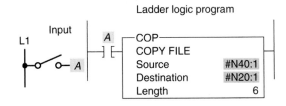

Figure 10-6 File copy (COP) rung for question 12.

12. For the file copy (COP) instruction rung of Figure 10-6,   12._____

a) both source and destinations are file addresses.

b) when input A goes true, the values in file N40 are copied to N20.

c) both file N40 and N20 contain 6 words.

d) all of these

13. Which of the following is *not* considered to be a data compare   13._____
instruction?

a) LESS THAN   c) MOV

b) EQUAL   d) GREATER THAN

Ladder logic program

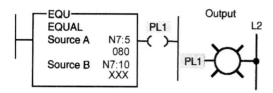

Figure 10-7 Logic rung for question 14.

14. Output PL1, of the logic rung shown in Figure 10-7, will be true when 14._____
the value of the number stored in word N7:10 is

a) less than 80.   c) equal to 80.

b) greater than 80.   d) equal to or greater than 80.

Ladder logic program

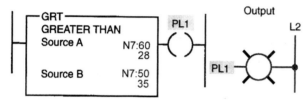

Figure 10-8 Logic rung for question 15.

15. The logic rung in Figure 10-8   15._____

a) has logic continuity.   c) will cause output PL1 to be energized.

b) does not have logic continuity.   d) both a and c

Ladder logic program

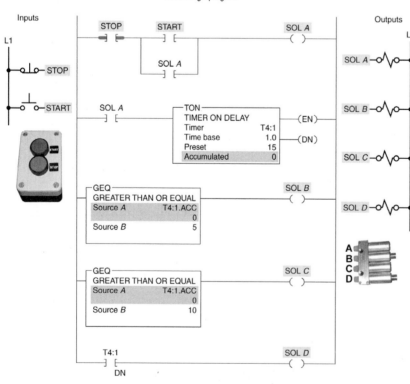

Figure 10-9 Timer program for question 16.

16-1. In the timer program of Figure 10-9, solenoid B is energized   16-1.____

a) as soon as the start button is pressed.   c) 10 s after the start button is pressed.

b) 5 s after the start button is pressed.   d) 15 s after the start button is pressed.

16-2. Source A of the GREATER THAN OR EQUAL instructions   16-2.____

contains the

a) preset time value of the timer.   c) reset time value of the timer.

b) time base value of the timer.   d) accumulated time value of the timer.

16-3. Bit T4:1/DN is called the   16-3.____

a) reset bit.   c) timing bit.

b) done bit.   d) data compare bit.

Ladder logic program

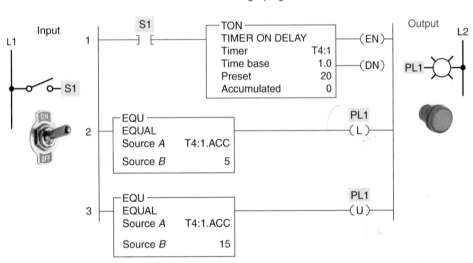

Figure 10-10 Timer program for question 17.

17-1. In the timer program of Figure 10-10, when switch S1 is closed   17-1.____

a) the light turns on after 20 s and remains on.

b) the light turns on after 5 s and remains on.

c) the light turns on after 15 s and remains on.

d) the light turns on after 5 s, stays on for 10 s, and then turns off.

17-2. Ten seconds immediately after switch S1 is closed, rung(s) ____   17-2.____

will be true.

a) No. 1 only   c) No. 1 and No. 3

b) No. 1 and No. 2   d) No. 1, No. 2, and No. 3

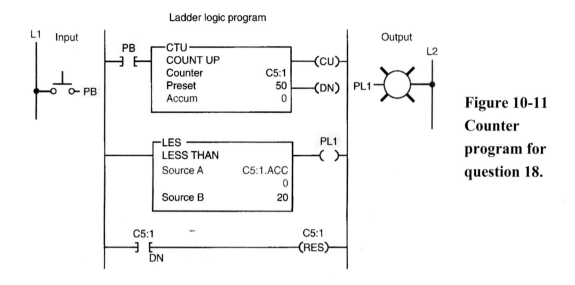

Ladder logic program

Figure 10-11 Counter program for question 18.

18-1. For the counter program of Figure 10-11, source A is addressed to the   18-1.____
a) pushbutton input.   c) done bit of the counter.
b) light output.   d) accumulated value of the counter.

18-2. Output PL1 will be true when the accumulated value of the counter is   18-2.____
a) equal to 50.   c) between 0 and 19.
b) equal to 20.   d) between 20 and 50.

18-3. The LESS THAN instruction is   18-3.____
a) always true.
b) true as long as the value contained in source A is less than 20.
c) true as long as the value contained in source A is more than 20.
d) true as long as the value contained in source A is equal to 20.

18-4. Assume that the input pushbutton is pulsed 60 times after the counter   18-4.____
had first been reset. After this operational sequence,
a) the light would be on and the accumulated count would be 10.
b) the light would be off and the accumulated count would be 10.
c) the light would be on and the accumulated count would be 60.
d) the light would be off and the accumulated count would be 60.

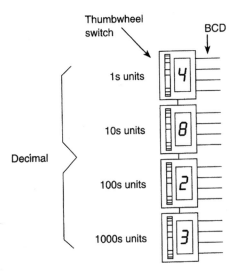

Figure 10-12 Thumbwheel switch for question 19.

19. The BCD value for the thumbwheel switch setting of Fig. 10-12 would be 19._____
a) 1010 0110 1110 1101.   c) 0011 0010 1000 0100.
b) 1110 0001 1010 1010.   d) 0010 1010 0110 1011.

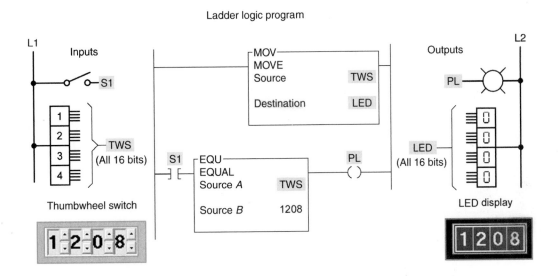

Ladder logic program

Figure 10-13 Program for question 20.

20-1 For the program of Figure 10-13, the thumbwheel switch is connected   20-1.____
to a ____ input module.
a) BCD   c) PID
b) discrete   d) LED

20-2. The MOV instruction is used to move data

20-2.____

a) from the LED display to the thumbwheel switch.

b) from the thumbwheel switch to the LED display.

c) from the LED display to the PL.

d) from the thumbwheel switch to the PL.

20-3. The PL output is energized when

20-3.____

a) S1 is closed and the value of the TWS is 1208.

b) S1 is open and the value of the TWS is 1208.

c) S1 is closed and the value of the TWS is less than 1208.

d) S1 is closed and the value of the TWS is greater than 1208.

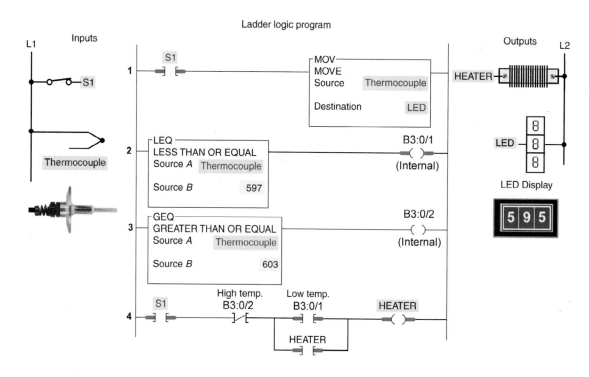

Figure 10-14 Temperature control program for question 21.

21-1. For the temperature control program of Figure 10-14, rung No. 1 contains the logic that

21-1.____

a) detects when the temperature drops below the low set point.

b) allows the thermocouple temperaure to be monitored by the LED display board.

c) detects when the temperature rises above the high set point.

d) switches the heater on and off.

21-2. Rung No. 2 contains the logic that

a) detects when the temperature drops below the low set point.

b) allows the thermocouple temperature to be monitored by the LED display board.

c) detects when the temperature rises above the high set point.

d) switches the heater on and off.

21-2.____

21-3. Rung No. 3 contains the logic that:

a) detects when the temperature drops below the low set point.

b) allows the thermocouple temperature to be monitored by the LED display board.

c) detects when the temperature rises above the high set point.

d) switches the heaters on and off.

21-3.____

21-4. Rung No. 4 contains the logic that

a) detects when the temperature drops below the low set point.

b) allows the thermocouple temperature to be monitored by the LED display board.

c) detects when the temperature rises above the high set point.

d) switches the heaters on and off.

21-4.____

22. In a closed-loop control system, the PLC control program acts to

a) form a closed circuit between input and output modules.

b) monitor the output signal and adjust the input signal accordingly.

c) keep the input and output in balance.

d) correct any difference between the measured value and the desired value.

22._____

23. The move instruction copies data from a(n) ____ word to a(n) __ word.

a) source, destination   c) integer, floating point

b) destination, source   d) floating point, integer

23._____

24. With the masked move instruction, where there is a ____ in the mask, data will pass.

a) 1   c) negative sign

b) 0   d) positive sign

24._____

25. The _____ mode of the FAL instruction allows one element of data to be operated on for every false-to-true transition of the instruction.

25._____

a) all
b) numeric
c) incremental
d) sequential

26. PID controllers produce outputs that that depend on

26._____

a) magnitude of the error signal.
b) duration of the error signal.
c) rate of change of the error signal.
d) all of these

27. The transfer of data from a word location to a file is called a

27._____

a) file-to-file move.
b) file-to-word move.
c) word-to-file move.
d) word-to-word move.

TEST 10.2

Place the answers to the following questions in the answer column at the right.

1. The source of a move instruction contains a copy of the data that
is to be moved. (True or False)

1._____

2. The ____ of a move instruction contains the address where the instruction
sends a copy of the data.

2._____

3. Each data manipulation instruction requires two or more ____ of data
memory for operation.

3._____

4. The words of data memory in singular form may be referred to as
either words or ____.

4._____

5. A consecutive group of data memory words may be referred to as
either a(n) (a) ____ or a(n) (b) ____.

5a._____
5b._____

6. The data contained in words will be in the form of binary ____
represented as series of 1s and 0s.

6._____

7. The format used for data manipulation instructions is the same for all
PLC models. (True or False)

7._____

8. Data manipulation can be placed into the two broad categories
of data (a) ____ and data (b) ____.

8a._____
8b._____

9. Data ____ instructions involve the transfer of the contents from one word
or register to another.

9._____

10. Data ____ instructions compare the data stored in two or more words.

10._____

11. Data transfer instructions can address only a limited number of special
locations in the memory. (True or False)

11._____

12. The MOV instruction is used to ____ the value in one word to another.

12._____

203

13. When the move instruction is true, the value stored at the (a) _____ address is copied into the (b) _____ address.

13a._____
13b._____

14. The move with mask instruction uses a mask to filter out _____ that are not to be transferred from the source to the destination.

14._____

15. Numerical data I/O interfaces are used to interface (a) _____ digital devices and (b) _____ devices.

15a._____
15b._____

16. Multibit interfaces allow a(n) _____ of bits to be input or output as a unit.

16._____

17. Multibit interfaces are used to accommodate devices that require BCD input or outputs. (True or False)

17._____

18. The analog input module contains a digital-to-analog converter circuit. (True or False)

18._____

19. An analog I/O will allow monitoring and control of _____ voltages and currents.

19._____

20. The analog output interface module receives numerical data from the processor that are translated into a proportional (a) _____ or (b) _____.

20a._____
20b._____

21. Set-point control in its simplest form _____ an input value to a set-point value.

21._____

22. Four types of set-point control are (a) _____, (b) _____, (c) _____, and (d) _____.

22a._____
22b._____
22c._____
22d._____

23. Each type of set-point control involves the use of some form of _____ loop control.

23._____

24. To copy the value in N12:0 into N12:40 using the MOV instruction, you would enter (a) _____ as the source and (b) _____ as the destination.

24a._____
24b._____

25. To put the value of 0 into N10:0 through N10:150 using the FLL instruction, you would enter (a) _____ as the source, (b) _____ as the destination, and (c) _____ as the length.

25a._____
25b._____
25c._____

26. To put the value in the upper half of N13:30 into the upper half of N12:0 using the MVM instruction, you would enter (a) _____ as the source, (b) _____ as the mask, and (c) _____ as the destination.

26a._____
26b._____
26c._____

```
┌─ EQU ──────────┐
│  EQUAL         │
│  Source A  N7:2│
│  Source B    25│
└────────────────┘
```
Figure 10-15 Instruction for question 27.

27. What value(s) stored in N7:2 of Figure 10-15 would make this instruction true?

27._____

```
┌─ NEQ ──────────┐
│  NOT EQUAL     │
│  Source A  N7:2│
│  Source B    20│
└────────────────┘
```
Figure 10-16 Instruction for question 28.

28. What value(s) stored in N7:2 of Figure 10-16 would make this instruction true?

28._____

```
┌─ LES ──────────┐
│  LESS THAN     │
│  Source A  N7:2│
│  Source B  N7:5│
│              10│
└────────────────┘
```
Figure 10-17 Instruction for question 29.

29. What value(s) stored in N7:2 of Figure 10-17 would make this instruction true?

29._____

```
┌─ GRT ──────────┐
│  GREATER THAN  │
│  Source A  N7:2│
│  Source B  N7:5│
│              10│
└────────────────┘
```
Figure 10-18 Instruction for question 30.

30. What value(s) stored in N7:2 of Figure 10-18 would make this instruction true?

30._____

```
┌─ GEQ ──────────────────────┐
│  GREATER THAN OR EQUAL TO   │
│  Source A            N7:2   │
│  Source B            N7:5   │
│                        10   │
└─────────────────────────────┘
```
Figure 10-19 Instruction for question 31.

31. What value(s) stored in N7:2 of Figure 10-19 would make this instruction true?   31._____

```
┌ LIM ──────────────┐
│ LIMIT TEST         │
─┤ Low limit    100   ├─     **Figure 10-20 Instruction for question 32.**
│ Test         N7:2  │
│ High limit    200  │
└────────────────────┘
```

32. What value(s) stored in N7:2 of Figure 10-20 would make this instruction true?   32._____

```
┌ LIM ──────────────┐
│ LIMIT TEST         │
─┤ Low limit    200   ├─     **Figure 10-21 Instruction for question 33.**
│ Test         N7:2  │
│ High limit    100  │
└────────────────────┘
```

33. What value(s) stored in N7:2 of Figure 10-21 would make this instruction true?   33._____

```
┌ MEQ ───────────────┐
│ MASKED EQUAL TO     │
─┤ Source             ├─
│                     │
│ Mask         B3:4   │          **Figure 10-22 Instruction for question 34.**
│                     │
│ Compare      B3:5   │
└─────────────────────┘
B3:4 = 0000 0000 0000 1111
B3:5 = 0000 0000 0000 0111
```

34. What value(s) stored in the source address of Figure 10-22 would make this instruction true?   34._____

35. The move instruction _____ a value from one location to another.   35._____

36. The bit distribute instruction is used to move _____ within a word or between words.   36._____

37. Files allow large amounts of data to be scanned quickly. (True or False)   37._____

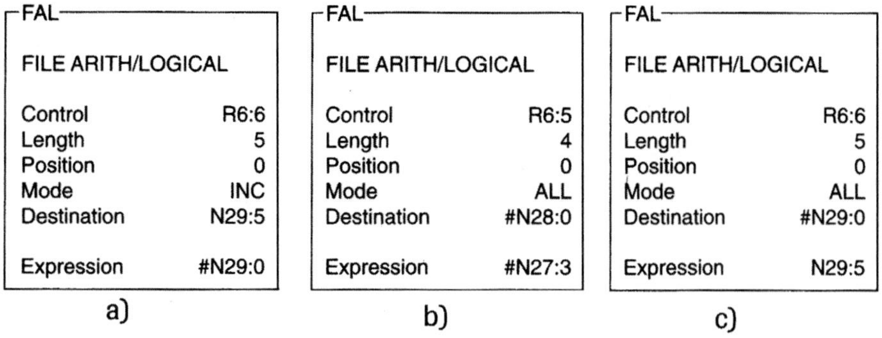

```
┌FAL─────────────────┐   ┌FAL─────────────────┐   ┌FAL─────────────────┐
│                    │   │                    │   │                    │
│ FILE ARITH/LOGICAL │   │ FILE ARITH/LOGICAL │   │ FILE ARITH/LOGICAL │
│                    │   │                    │   │                    │
│ Control      R6:6  │   │ Control      R6:5  │   │ Control      R6:6  │
│ Length          5  │   │ Length          4  │   │ Length          5  │
│ Position        0  │   │ Position        0  │   │ Position        0  │
│ Mode          INC  │   │ Mode          ALL  │   │ Mode          ALL  │
│ Destination  N29:5 │   │ Destination #N28:0 │   │ Destination #N29:0 │
│                    │   │                    │   │                    │
│ Expression  #N29:0 │   │ Expression  #N27:3 │   │ Expression   N29:5 │
└────────────────────┘   └────────────────────┘   └────────────────────┘
          a)                       b)                       c)
```

Figure 10-23 FAL instructions for question 38.

38. For each instruction shown Figure 10-23, signify the type of FAL copy operation used (file to file, file to word, or word to file).

35a._____
35b._____
35c._____

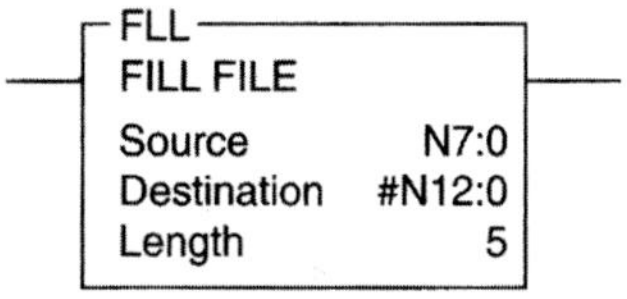

```
┌FLL──────────────┐
│  FILL FILE       │
│  Source     N7:0 │
│  Destination #N12:0│
│  Length        5 │
└──────────────────┘
```

Figure 10-24 FLL instructions for question 39.

39. The FLL instruction shown in Figure 10-24, when true, tells the processor to (a) _____ the value of word N7:0 into the first (b) _____ words of (c) _____ N12:0.

39a._____
39b._____
39c._____

40. The COP instruction operates at a lower speed than the same operation that uses the FAL instruction. (True or False)

40._____

41. The FLL instruction is frequently used to zero all the data in a file. (True or False)

41._____

42. Data transfer and data compare instructions are both output instructions. (True or False)

42._____

43. Input and output modules can be addressed either at the (a) _____ level or at the (b) _____ level.

43a._____
43b._____

44. PID control is inexpensive but not accurate enough for many applications. (True or False)

44._____

Programming Assignments

This section requires you to simulate several data manipulation applications. The instructions used are intended to be generic in nature and, as such, will require some conversion for the particular PLC model you are using. The use of a prewired PLC input/output control panel is recommended to simulate the operation of these circuits.

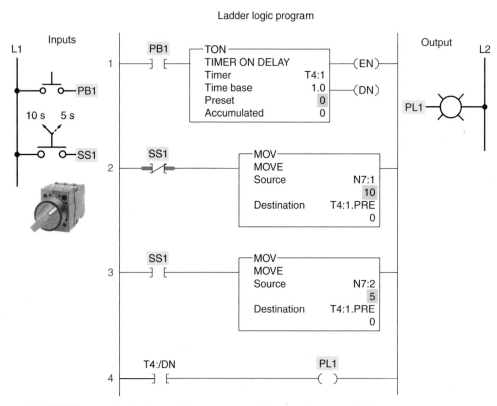

Figure 10-25 Timer data transfer program for assignment 1.

1) The timer data transfer program of Figure 10-25 is described in the text. Prepare an I/O connection diagram and ladder logic program that will simulate its operation. Use addresses that apply to your PLC installation. Enter the program into the PLC, and verify its operation.

Ladder logic program

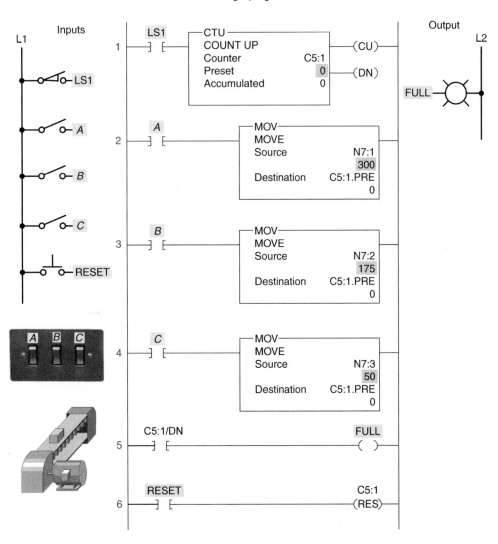

Figure 10-26 Counter data transfer program for assignment 2.

2) The counter data transfer program of Figure 10-26 is described in the text. Prepare an I/O connection diagram and ladder logic program that will simulate its operation. Use addresses that apply to your PLC installation. Enter the program into the PLC, and verify its operation.

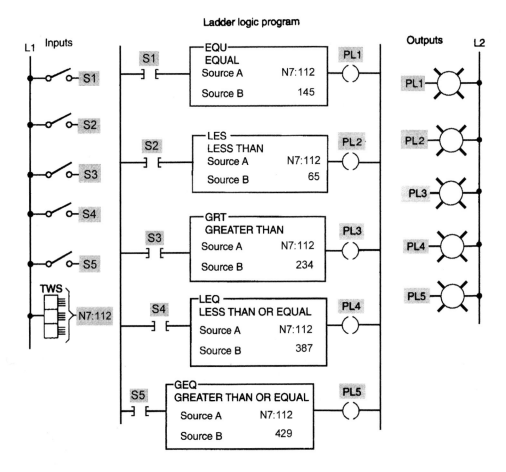

Figure 10-27 Data compare program for assignment 3.

3) Construct the data compare program of Figure 10-27 using a thumbwheel switch interface module for the changing variable. Enter the program into the PLC, and prove the operation of each rung.

Ladder logic program

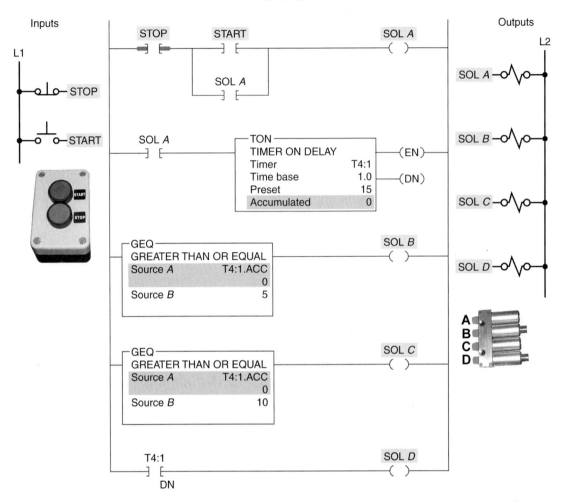

Figure 10-28 Timer program for assignment 4.

4) A timer program controlling multiple loads using a single timer and the GEQ instruction is shown in Figure 10-28 and described in the text. Prepare an I/O connection diagram and ladder logic program that will simulate its operation. Use addresses that apply to your PLC installation. Enter the program into the PLC, and verify its operation.

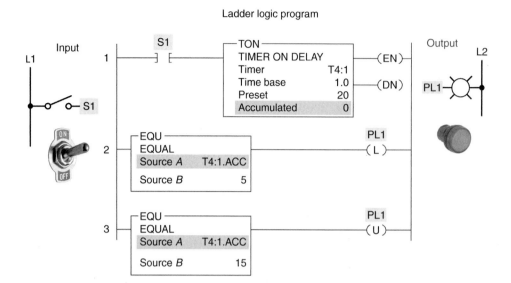

Figure 10-29 On-delay timer program for assignment 5.

5) An on-delay timer program implemented using the EQU instruction is shown in Figure 10-29 and described in the text. Prepare an I/O connection diagram and ladder logic program that will simulate its operation. Use addresses that apply to your PLC installation. Enter the program into the PLC, and verify its operation.

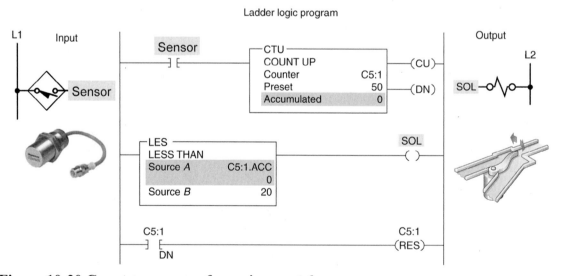

Figure 10-30 Counter program for assignment 6.

6) A counter program implemented using the LES instruction is shown in Figure 10-30 and described in the text. Prepare an I/O connection diagram and ladder logic program that will simulate its operation. Use addresses that apply to your PLC installation. Enter the program into the PLC, and verify its operation.

Ladder logic program

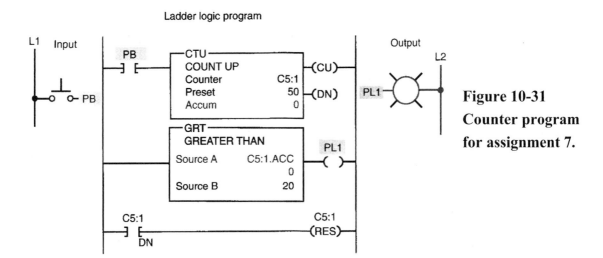

Figure 10-31 Counter program for assignment 7.

7) Construct the counter program shown in Figure 10-31 using your PLC demonstration panel. After constructing the program on paper, enter it into the PLC. Demonstrate that the output will be energized when the counter's accumulated value is from 21 to 50 and that the counter will reset automatically when it reaches its preset value of 50.

Ladder logic program

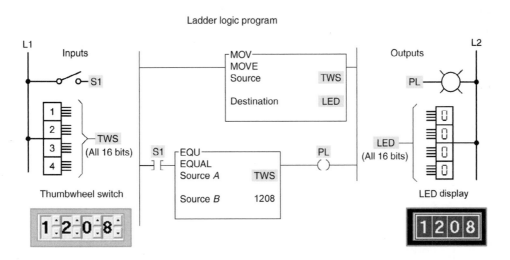

Figure 10-32 Thumbwheel switch program for assignment 8.

8) Monitoring the setting of a thumbwheel switch program of Figure 10-32 is described in the text. Prepare an I/O connection diagram and ladder logic program that will simulate its operation. Use addresses that apply to your PLC installation. Enter the program into the PLC, and demonstrate that the decimal setting of the thumbwheel switches is monitored by the LED display board and that pilot light PL will turn on when switch S1 is closed and the value of the thumbwheel switches is 1208.

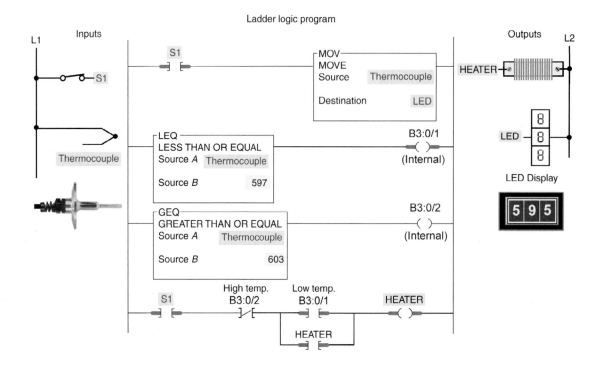

Figure 10-33 Set-point control switch program for assignment 9.

9a) The set-point control program of Figure 10-33 is described in the text. Prepare an I/O connection diagram and ladder logic program that will simulate its operation. Use addresses that apply to your PLC installation. Enter the program into the PLC, and verify its operation.

  b) Modify the program to include alarm lights that come on if the temperature rises above 610°F or drops below 590°F. Once on, each alarm light stays on until manually reset with a pushbutton.

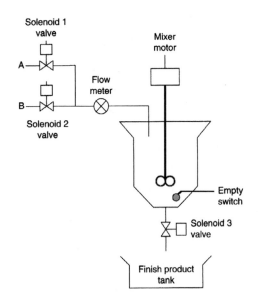

Figure 10-34 Process flow diagram for assignment 10.

10) Design a PLC program to implement the process of mixing two ingredients, A and B, as illustrated in Figure 10-34. The mixing cycle can be summarized as follows:

- Ingredient A is sent to the tank first by energizing solenoid No. 1. The flow meter gives one pulse for every gallon of flow. Solenoid valve No. 1 will be open (energized) until 200 gallons have poured in.
- After ingredient A is in the tank, 300 gallons of ingredient B should be added. The process of adding ingredient B follows the same procedure as that of adding ingredient A.
- After ingredient B is in the tank, the mixer motor starts and runs for 5 minutes.
- After the mixing is complete, solenoid No. 3 should open and let the mixed batch go into a finished tank.
- When the tank is empty (as indicated by the NC empty liquid-level switch), solenoid No. 3 should close and stop the cycle.

Enter the simulated program into the PLC, and prove its operation.

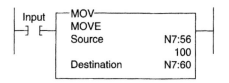

Figure 10-35 Move instruction program for assignment 11.

11) Enter the move instruction program of Figure 10-35 into the PLC, and prove its operation.

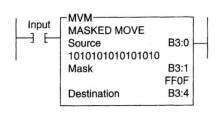

Figure 10-36 Masked move instruction program for assignment 12.

12) Enter the masked move instruction program of Figure 10-36 into the PLC, and prove its operation.

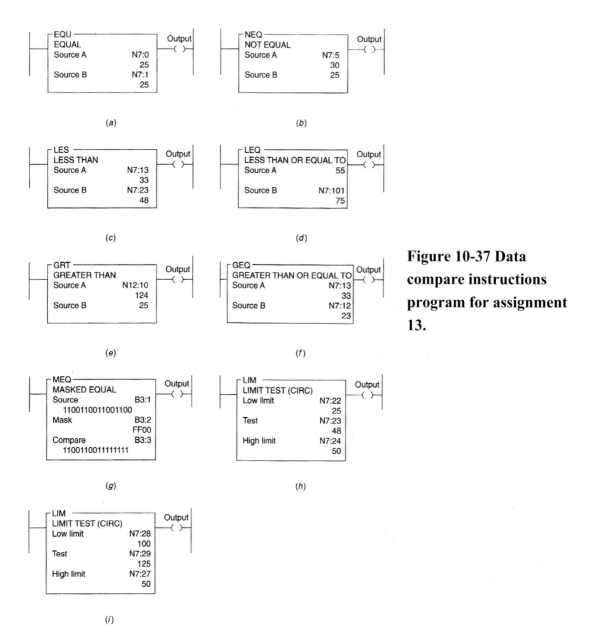

Figure 10-37 Data compare instructions program for assignment 13.

13) Enter each of the data comparison instructions, (a) through (i), shown in Figure 10-37 individually into the PLC, and prove the operation of each.

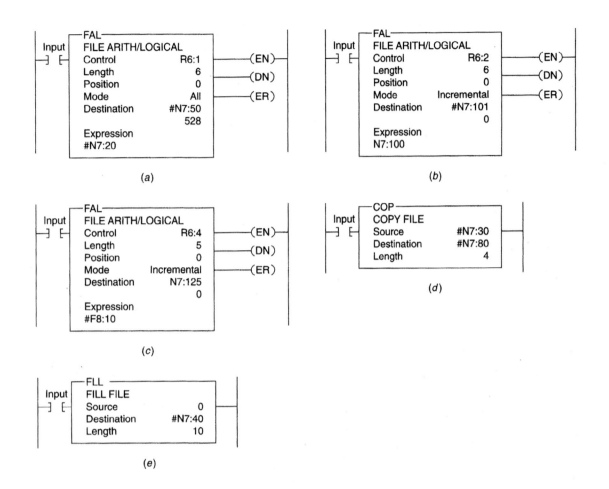

Figure 10-38 Copy instructions program for assignment 14.

14) Enter each of the file copy instructions, (a) through (e), shown in Figure 10-38 individually into the PLC, and prove the operation of each.

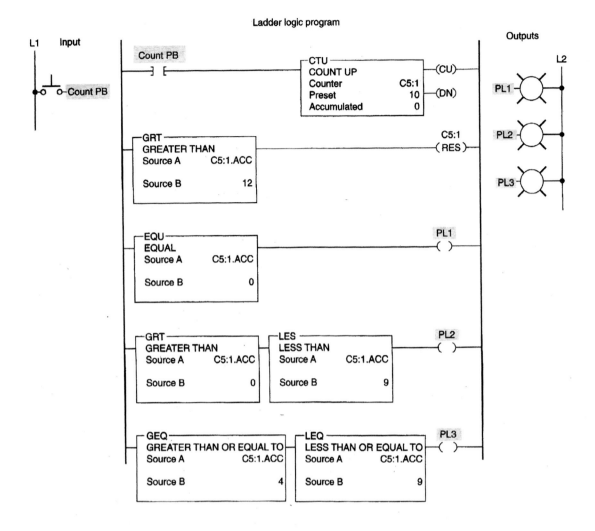

Figure 10-39 Data comparison program for assignment 15.

15) Enter the data comparison program of Figure 10-39 into the PLC. Operate the program, and answer the following questions about the operation:

  a) What is the highest count achieved before the counter is reset?

  b) At what accumulated values of the counter are lights PL1, PL2, and PL3 energized?

16) Construct and test each of the following PLC data compare problems:

  a) A light is to come on only if a PLC counter has an accumulated value of 8 or 14.

  b) A light is to be on if a PLC counter does not have accumulated values of either 6 or 10.

  c) A light is to come on if three PLC counters have the same accumulated values.

17) Design and test a PLC program to implement a solution to the following problem:

A room heating and air conditioning system is to be implemented with a programmable controller. The room temperature is read by a temperature transducer to word N33:1. The outdoor temperature is read by another temperature transducer to word N33:2. The logic to place these temperatures into these locations is assumed to be already in place. Design logic to:

- Turn on the heat when the indoor temperature is at or below 21°C and the outdoor temperature is below 16°C.
- Turn off the heat when the temperature is at or above 22°C.
- Turn on the air conditioning when the indoor temperature is at or above 22°C and the outdoor temperature is above 20°C.
- Turn off the air conditioning when the indoor temperature is at or below 21°C.

18) A baking process includes three ovens (No. 1, No. 2, and No. 3), each controlled by a separate PLC timer. The baked product is to remain in each oven for a specified time according to the recipe produced. There are three separate recipes to run through the ovens. The following gives the bake time, in seconds, for each recipe:

OVEN BAKE TIMES

Recipe	No. 1	No. 2	No. 3
A	10 s	20 s	5 s
B	8 s	12 s	48 s
C	24 s	16 s	4 s

Construct a program that will allow an operator to select and run any one of the three recipes. Enter the program into the PLC, and prove its operation.

TEST 11.1

Choose the letter that best completes the statement.

1. The ability of a PLC to perform math functions is intended to   1._____
a) replace a calculator.
b) multiply the effective number of input and output devices.
c) perform arithmetic functions on values stored in memory words.
d) all of these

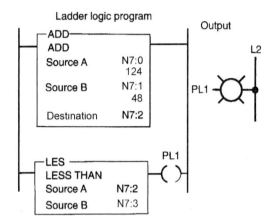

Figure 11-1 Program for question 2.

2-1. In the program of Figure 11-1, the value of the number stored in N7:2 is   2-1._____
a) 172.   c) 325.
b) 601.   d) 348.

2-2. Which of the following numbers stored in N7:3 will cause output PL1   2-2._____
to be energized?
a) 048   c) 172
b) 124   d) 325

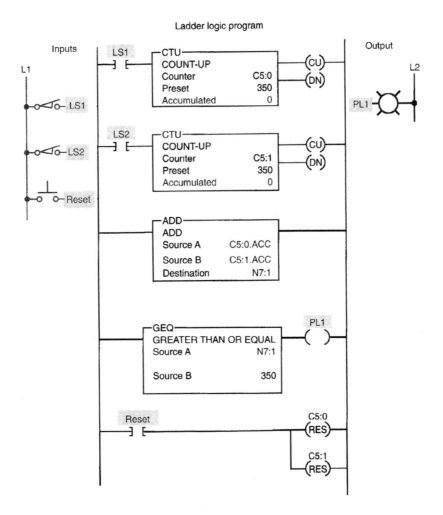

Figure 11-2 Counter program for question 3.

3-1. In the counter program of Figure 11-2, assume the accumulated count   3-1._____
of counters C5:0 and C5:1 to be 124 and 248, respectively. As a result,
a) the number 372 will be stored in word N7:1 and output PL1 will be energized.
b) the number 372 will be stored in word N7:1 and output PL1 will not be energized.
c) the number 350 will be stored in word N7:1 and output PL1 will be energized.
d) the number 350 will be stored in word N7:1 and output PL1 will not be energized.

3-2. Assume that the light is to come on after a total count of 120. As a   3-2._____
result,
a) the preset counter C5:0 must be changed to 120.
b) the value in source B of the GEQ instruction must be changed to 120.
c) the value in source B of the ADD instruction must be changed to 120.
d) the value in word N7:1 must be changed to 120.

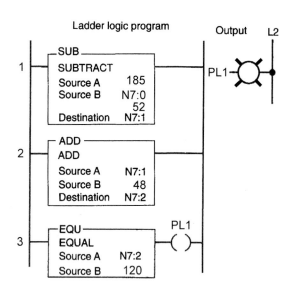

Figure 11-3 Program for question 4.

4-1. For the program of Figure 11-3, the number stored in N7:2 would be   4-1._____

a) 85.   c) 181.

b) 28.   d) 285.

4-2. Rung No. 2 will be true   4-2._____

a) at all times.

b) when the number stored in word N7:1 is equal to 48.

c) when the number stored in word N7:1 is less than 48.

d) when the number stored in word N7:1 is greater than 48.

4-3. Output PL1   4-3._____

a) would be energized.

b) would not be energized.

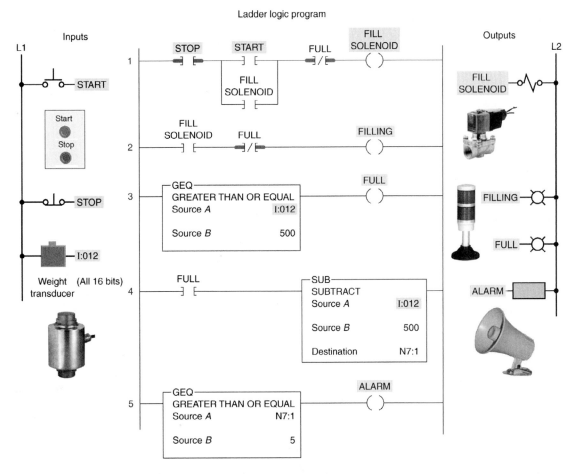

Figure 11-4 Vessel filling program for question 5.

5-1. For the program of Figure 11-4, the preset full weight of the vessel   5-1._____
is changed by changing

a) the value of the number stored at input I:012.

b) the value of source B of the GEQ instruction of Rung 3.

c) the value of source B of the GEQ instruction of Rung 5.

d) the value of the number stored in word N7:1.

5-2. The amount of overfill weight required to trigger the alarm is changed   5-2._____
by changing

a) the value of the number stored at input I:012.

b) the value of source B of the GEQ instruction of Rung 3.

c) the value of source B of the GEQ instruction of Rung 5.

d) the value of the number stored in word N7:1.

5-3. When the Full light is on   5-3._____

a) the weight of the vessel is 500 pounds or more.

b) Rung No. 3 is always true.

c) Rung No. 1 is always false.

d) all of these

5-4. When the Filling light is on   5-4._____

a) the weight of the vessel is less than 500 pounds.

b) Rung No. 2 is always true.

c) Rung No. 4 is always false.

d) all of these

5-5. The number stored in word N7:1 represents the   5-5._____

a) weight of the empty vessel.

b) preset weight of the vessel.

c) current weight of the vessel.

d) difference between the current and preset weight of the vessel.

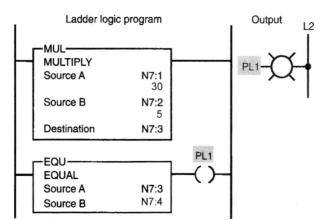

Figure 11-5 Program for question 6.

6-1. For the program of Figure 11-5, the number stored in N7:3 is   6-1._____

a) 6.   c) 150.

b) 60.   d) 300.

6-2. What number stored in N7:3 will turn PL1 on?   6-2._____

a) 150   c) 50

b) 100   d) All of these

224

6-3. Assume the value stored in N7:1 changes from 30 to 10. Which value   6-3._____
stored in N7:4 will result in PL1 being energized?

a) 10   c) 35

b) 25   d) 50

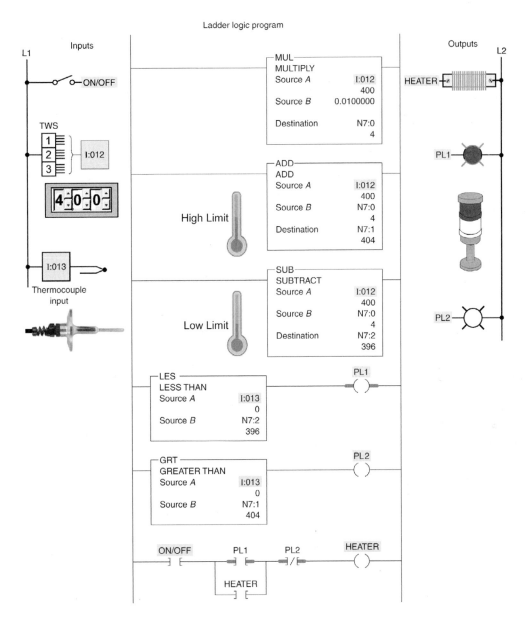

Ladder logic program

Figure 11-6 Temperature control program for question 7.

7-1. For the program of Figure 11-6, the set-point temperature is set   7-1._____
by the number stored in

a) N7:0.   c) I:012.

b) N7:2.   d) I:013.

7-2. The number stored in N7:1 represents the   7-2._____

a) upper temperature limit.

b) lower temperature limit.

c) current temperature of the oven.

d) difference between the preset and current temperature.

7-3. PL1 will be on whenever the current temperature is   7-3._____

a) greater than the preset temperature.

b) less than the preset temperature.

c) greater than the upper temperature limit.

d) less than the lower temperature limit.

7-4. The ADD instruction is telling the processor to add the   7-4._____

a) preset and current temperatures.

b) upper and lower temperature limits.

c) current and upper limit temperatures.

d) preset and upper deadband range.

7-5. Assume the set-point temperature is changed to 200°F. As a result, the   7-5._____
number stored in N7:0 would be

a) 2.   c) 6.

b) 4.   d) 8.

7-6. Assume the upper and lower temperature limits are programmed for 2%   7-6._____
instead of 1% and the preset is 400°F. As a result, the number stored in
N7:2 would be

a) 392.   c) 388.

b) 390.   d) 386.

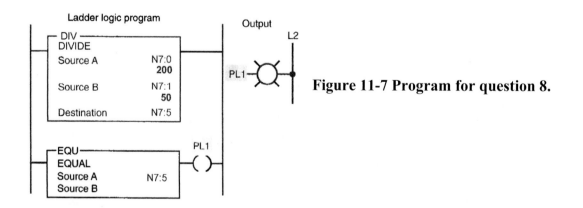

Figure 11-7 Program for question 8.

8-1. For the program of Figure 11-7, the number stored in N7:5 would be   8-1._____

a) 1000.   c) 4.

b) 500.   d) 2.

8-2. Which constant stored in source B of the EQU instruction would   8-2._____

turn PL1 on?

a) 24   c) 6

b) 20   d) 4

8-3. Assume the value stored in N7:0 is 90, the value stored at N7:1 is   8-3._____

3, and the constant for source B of the EQU instruction is 10. What would the

state of PL1 be?

a) Off   b) On

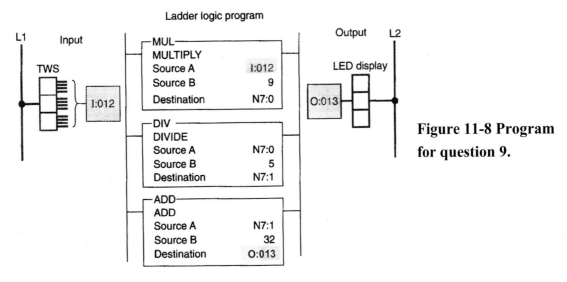

Figure 11-8 Program for question 9.

9-1. The program of Figure 11-8 is used to convert the Celsius temperature   9-1._____

indicated by the thumbwheel switch to Fahrenheit values for display.

Answer each of the questions with reference to this program, assuming a

thumbwheel switch setting of 25°C. The value of the number stored in I:012 is

a) 25.   c) 35.

b) 30.   d) 40.

9-2. The value of the number stored in N7:0 is   9-2._____

a) 225.   c) 750.

b) 500.   d) 230.

9-3. The value of the number stored in N7:1 is

a) 90.   c) 35.

b) 45.   d) 60.

9-3._____

9-4. The value of the number stored in O:013 is

a) 98.   c) 67.

b) 77.   d) 57.

9-4._____

10. Math instructions are all _____ instructions.

a) output   c) binary

b) input   d) BCD

10._____

11. File arithmetic functions are used to perform arithmetic operations on

a) multiple words.   c) decimal numbers only.

b) integer numbers only.   d) BCD numbers only.

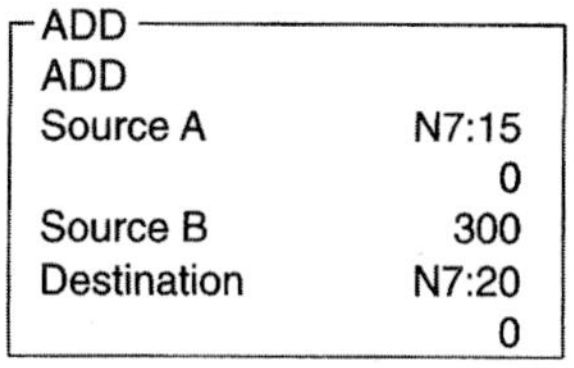

11._____

```
ADD
 ADD
 Source A          N7:15
                       0
 Source B            300
 Destination       N7:20
                       0
```

Figure 11-9 ADD instruction for question 12.

12. For the ADD instruction of Figure 11-9, the _____ used as part of the expression is a constant.

a) Source A   c) 300

b) 0   d) N7:20

12._____

13. Which math instruction would you use if you wanted to take the opposite sign of a value?

a) SUB   c) NEG

b) SQR   d) CLR

13._____

14. Which math instruction would you use if you wanted to calculate the difference between the accumulated values of two counters?

a) ADD   c) MUL

b) SUB   d) DIV

14._____

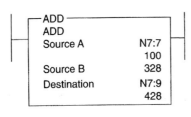

Figure 11-10 ADD instruction for question 15.

15. With reference to the ADD instruction of Figure 11-10, the value   15._____
of the number stored at Source B is

a) N7:8.

b) N7:16.

c) 328.

d) 528.

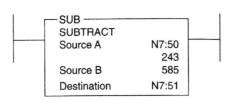

Figure 11-11 SUB instruction for question 16.

16. With reference to the SUB instruction of Figure 11-11, the value of the   16._____
number stored at Destination is

a) 293.

b) –193.

c) 51.

d) –342.

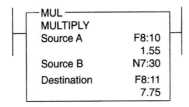

Figure 11-12 MUL instruction for question 17.

17. With reference to the MUL instruction of Figure 11-12, the value of the   17._____
number stored at N7:30 is

a) 5.

b) 15.5.

c) 4.87.

d) 1.85.

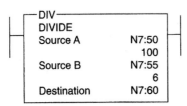

Figure 11-13 DIV instruction for question 18.

18. With reference to the DIV instruction of Figure 11-13, the value of the number stored at Destination is

a) 17.

b) 16.666.

c) 50.

d) 51.

18._____

TEST 11.2

Place the answers to the following questions in the answer column at the right.

1. Math instructions enable the programmable controller to take on some of the qualities of a _____ system.

1._____

2. The ability of a PLC to perform math functions is intended to allow it to replace a calculator. (True or False)

2._____

3. PLC math functions perform arithmetic on _____ stored in memory words.

3._____

4. The four basic math functions performed by PLCs are (a) _____, (b) _____, (c) _____, and (d) _____.

4a._____
4b._____
4c._____
4d._____

5. All PLC manufacturers use the same format for math instructions. (True or False)

5._____

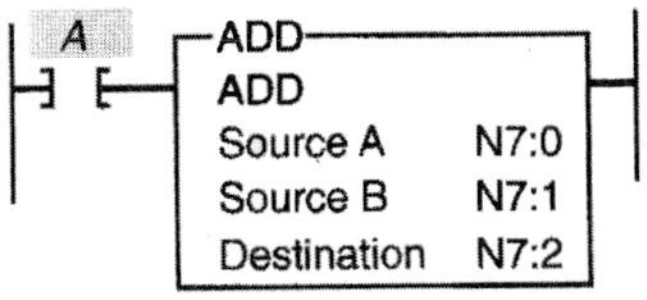

Figure 11-14 Logic rung for question 6.

```
   A    ┌ADD──────────────┐
─┤ ├────┤ ADD             ├─
        │ Source A   N7:0 │
        │ Source B   N7:1 │
        │ Destination N7:2│
        └─────────────────┘
```

6. The rung of Figure 11-14 is telling the processor to add the values stored in words (a) ___ ___ store the sum in word (b) _____ and whenever (c) _____ is true.

6a._____
6b._____
6c._____

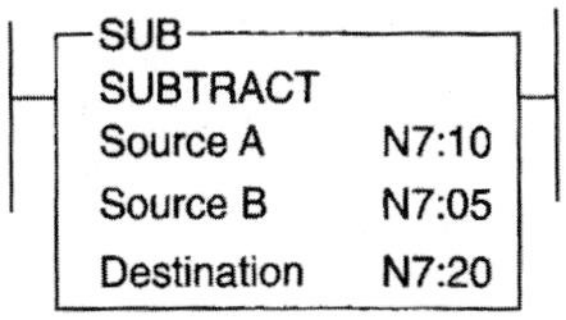

Figure 11-15 Logic rung for question 7.

```
   ┌SUB───────────────┐
───┤ SUBTRACT         ├
   │ Source A   N7:10 │
   │ Source B   N7:05 │
   │ Destination N7:20│
   └──────────────────┘
```

7. The rung of Figure 11-15 is telling the processor to subtract the value stored in word (a) _____ from the value stored in word (b) _____ and store the difference in word (c) _____.

7a._____
7b._____
7c._____

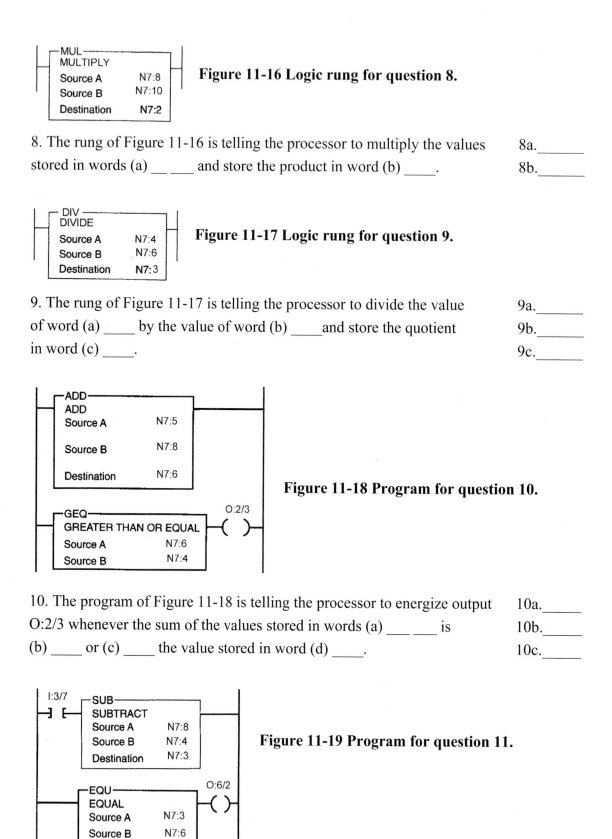

Figure 11-16 Logic rung for question 8.

8. The rung of Figure 11-16 is telling the processor to multiply the values stored in words (a) _____ and store the product in word (b) ____.

8a._____
8b._____

Figure 11-17 Logic rung for question 9.

9. The rung of Figure 11-17 is telling the processor to divide the value of word (a) ____ by the value of word (b) ____ and store the quotient in word (c) ____.

9a._____
9b._____
9c._____

Figure 11-18 Program for question 10.

10. The program of Figure 11-18 is telling the processor to energize output O:2/3 whenever the sum of the values stored in words (a) _____ is (b) ____ or (c) ____ the value stored in word (d) ____.

10a._____
10b._____
10c._____

Figure 11-19 Program for question 11.

11. The program of Figure 11-19 is telling the processor to energize output   11a._____
O:6/2 whenever (a) ____ is true and the difference between the values stored   11b._____
in words (b) ___ ___ is equal to the value stored in word (c) ____.   11c._____

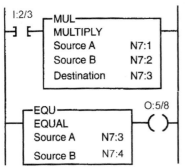

Figure 11-20 Program for question 12.

12. For the program of Figure 11-20, if output O:5/8 is to be energized   12._____
when the product of the values stored in words N7:1 and N7:2 is equal
to 1520, then the value of the number stored in word N7:4 must be ____.

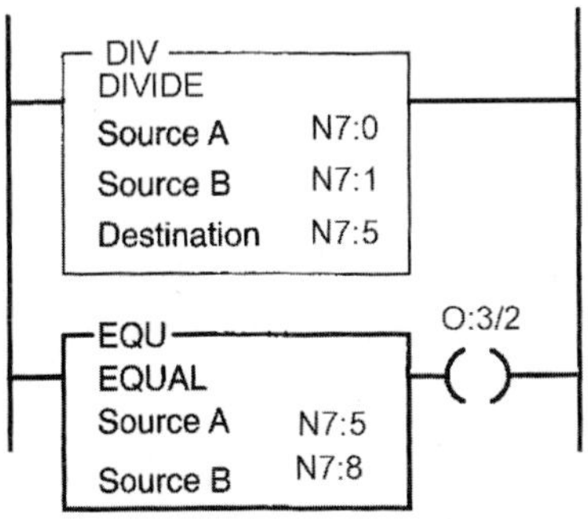

Figure 11-21 Program for question 13.

13. For the program of Figure 11-21, assume output O:3/2 is energized and   13._____
the values of the numbers stored in words N7:0 and N7:1 are 500 and 40,
respectively. The value of the number stored in word N7:8 would be ____.

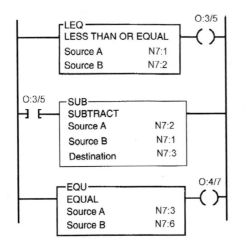

Figure 11-22 Program for question 14.

14. For the program of Figure 11-22, assume the value of the numbers stored in words N7:1, N7:2, and N7:6 are 600, 750, and 100, respectively. As a result, output O:3/5 state will be (a) ____, the number stored in word N7:3 will be (b) ____, and output O:4/7 state will be (c) ____.

14a._____
14b._____
14c._____

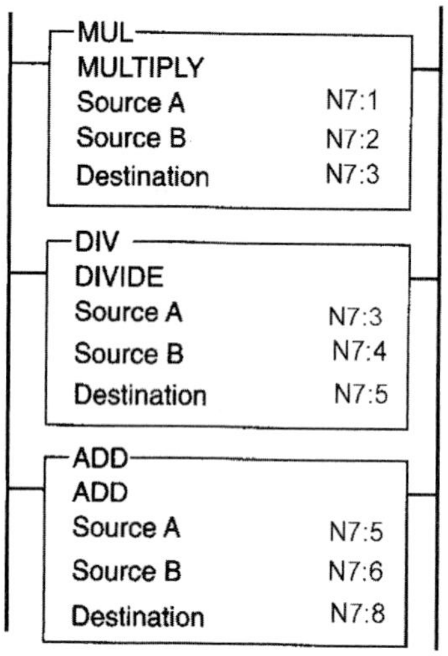

```
┌─MUL─────────────┐
│ MULTIPLY        │
│ Source A   N7:1 │
│ Source B   N7:2 │
│ Destination N7:3│
└─────────────────┘
┌─DIV─────────────┐
│ DIVIDE          │
│ Source A   N7:3 │
│ Source B   N7:4 │
│ Destination N7:5│
└─────────────────┘
┌─ADD─────────────┐
│ ADD             │
│ Source A   N7:5 │
│ Source B   N7:6 │
│ Destination N7:8│
└─────────────────┘
```

Figure 11-23 Program for question 15.

15. For the program of Figure 11-23, assume the value of the numbers stored in words N7:1, N7:2, N7:4, and N7:6, are 40, 9, 5, and 32, respectively. As a result, the value of the number stored in N7:3 is (a) ____ , in N7:5 is (b) ____, and in N7:8 is (c) ____.

15a._____
15b._____
15c._____

16. Math instructions are all output instructions. (True or False)

16._____

17. There is no limit to the maximum value a PLC math function can store. (True or False)

17._____

18. File arithmetic instructions are designed to perform math operations on single words. (True or False)

18._____

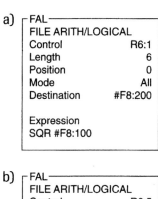

a)
```
┌─FAL──────────────────────┐
│ FILE ARITH/LOGICAL        │
│ Control            R6:1   │
│ Length                6   │
│ Position              0   │
│ Mode                All   │
│ Destination      #F8:200  │
│                           │
│ Expression                │
│ SQR #F8:100               │
└───────────────────────────┘
```

b)
```
┌─FAL──────────────────────┐
│ FILE ARITH/LOGICAL        │
│ Control            R6:5   │
│ Length                4   │
│ Position              0   │
│ Mode                  2   │
│ Destination      #N7:255  │
│                           │
│ Expression                │
│ #N10:0 − 255              │
└───────────────────────────┘
```

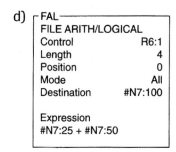

c)
```
┌─FAL──────────────────────┐
│ FILE ARITH/LOGICAL        │
│ Control            R6:8   │
│ Length                4   │
│ Position              0   │
│ Mode                All   │
│ Destination      #N7:500  │
│                           │
│ Expression                │
│ #N7:330 * N7:23           │
└───────────────────────────┘
```

d)
```
┌─FAL──────────────────────┐
│ FILE ARITH/LOGICAL        │
│ Control            R6:1   │
│ Length                4   │
│ Position              0   │
│ Mode                All   │
│ Destination      #N7:100  │
│                           │
│ Expression                │
│ #N7:25 + #N7:50           │
└───────────────────────────┘
```

Figure 11-24 FAL instructions for question 19.

e)
```
┌─FAL──────────────────────┐
│ FILE ARITH/LOGICAL        │
│ Control            R6:7   │
│ Length                4   │
│ Position              1   │
│ Mode        Incremental   │
│ Destination      F8:200   │
│                     0.1   │
│ Expression                │
│ #F8:20 I #F8:100          │
└───────────────────────────┘
```

19. Identify the math function for each of the FAL instructions shown in Figure 11-24.

19a._____

19b._____

19c._____

19d._____

19e._____

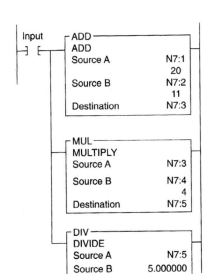

Figure 11-25 Program for question 20.

235

20. With reference to the program of Figure 11-25, when the input goes true, determine the value that will be stored in each of the following words.

20a._____
20b._____
20c._____

a) N7:3

c) F8:1

b) N7:5

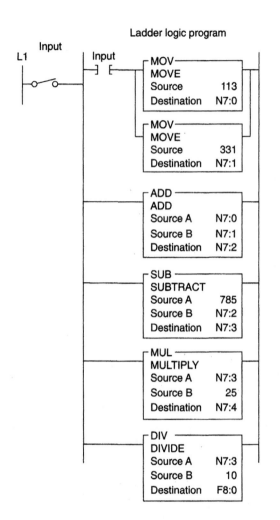

Ladder logic program

Figure 11-26 Program for question 21.

21. With reference to the program of Figure 11-26, when the input goes true, determine the value that will be stored in each of the following words.

a) N7:0   21a._____
b) N7:1   21b._____
c) N7:2   21c._____
d) N7:3   21d._____
e) N7:4   21e._____
f) F8:0   21f._____

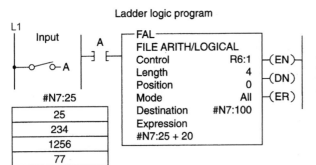

Ladder logic program

Figure 11-27 Program for question 22.

22. With reference to the program of Figure 11-27, when the input goes true, determine the value that will be stored in each of the following words.

a) N7:100   22a._____

b) N7:101   22b._____

c) N7:102   22c._____

d) N7:103   22d._____

Programming Assignments

This section requires you to simulate several arithmetic functions. The math instructions used are intended to be generic in nature and, as such, will require some conversion for the particular PLC model you are using. The use of a prewired PLC input/output control panel is recommended to simulate the operation of these circuits.

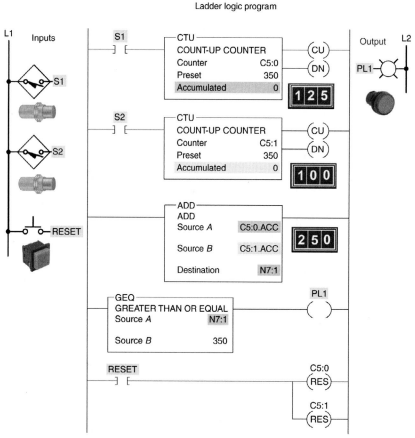

Figure 11-28 Counter program for assignment 1.

1) The counter program of Figure 11-28 is described in the text.

a) Prepare an I/O connection diagram and ladder logic program that will simulate its operation. Use addresses that apply to your PLC installation. Enter the program into the PLC, and verify its operation.

b) Modify the program so that a second light comes on when the accumulated count of the two counters is equal to 345 and remains on until the reset button is pressed. Enter the modified program into the PLC, and prove its operation.

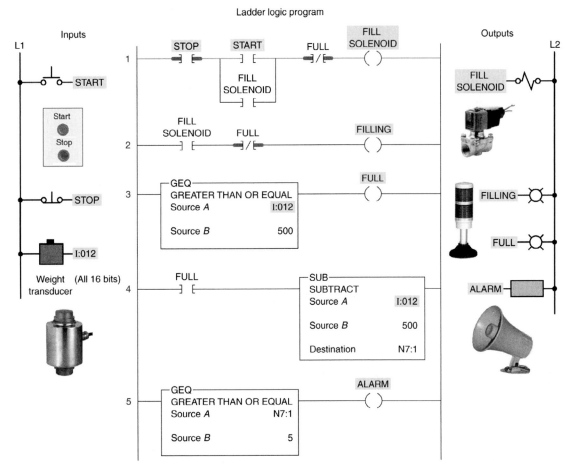

Figure 11-29 Vessel overfill alarm program for assignment 2.

2) The vessel overfill alarm program of Figure 11-29 is described in the text.
a) Prepare an I/O connection diagram and ladder logic program that will simulate its operation. Use addresses that apply to your PLC installation. Enter the program into the PLC, and verify its operation.
b) Modify the program so that should an overfill condition of 5 pounds or more occurs, an overfill solenoid is energized to automatically reduce the level back down to the 500-pound point. Enter the modified program into the PLC and prove its operation.

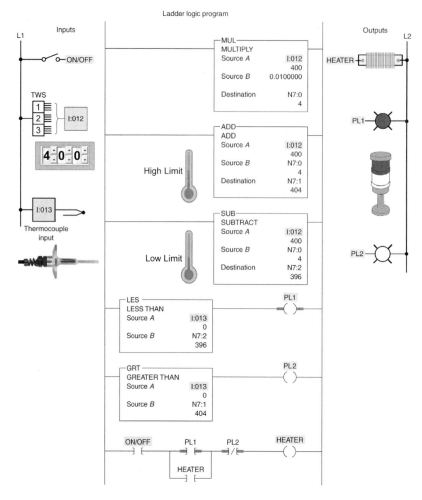

Figure 11-30 Temperature control program for assignment 3.

3) The temperature control program of Figure 11-30 is described in the text.

 a) Prepare an I/O connection diagram and ladder logic program that will simulate its operation. Use addresses that apply to your PLC installation. Enter the program into the PLC, and verify its operation.

 b) Modify the program to include each of the following:

- An LED output module to display the actual temperature
- A high temperature light to come on if the temperature rises above 410°F
- A low temperature light to come on if the temperature drops below 390°F

Enter the modified program into the PLC, and prove its operation.

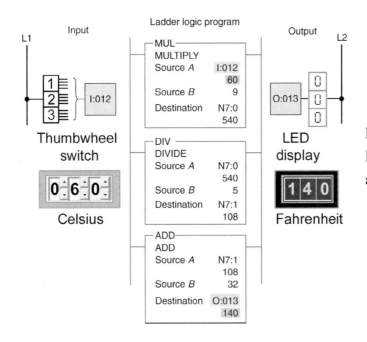

Figure 11-31 Celsius to Fahrenheit program for assignment 4.

4) The Celsius to Fahrenheit conversion program of Figure 11-31 is described in the text. Prepare an I/O connection diagram and ladder logic program that will simulate its operation. Use addresses that apply to your PLC installation. Enter the program into the PLC and verify its operation.

5) Design a simulated PLC program that will control the temperature of a furnace and monitor the temperature between 87°C and 100°C. An analog thermocouple input that measures Celsius temperature is to be used. The operation of the program can be summarized as follows:

- The sensed Celsius temperature is to be converted to Fahrenheit for display.
- When the displayed temperature drops below 190°F for a minimum of 5 seconds, a heater is turned on to bring the temperature back into the desired range. The heater stays on until the temperature rises back to 190°F.
- Should the displayed temperature reach 212°F, an alarm is turned on and remains on until manually reset with a pushbutton.

After constructing your program on a separate sheet of paper, enter it into the PLC and prove its operation.

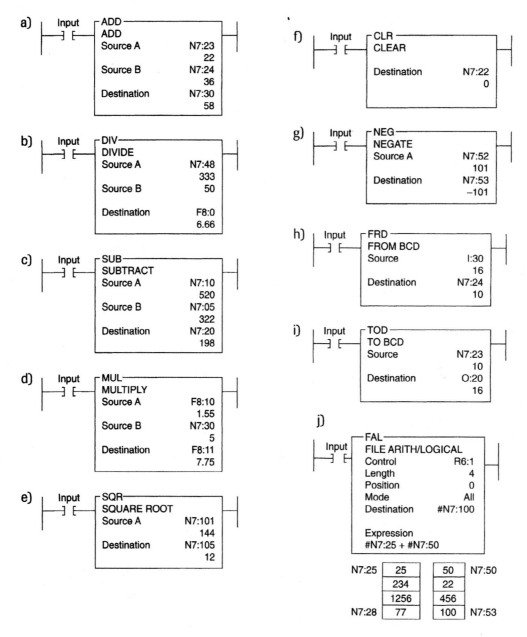

Figure 11-32 Math operations for assignment 6.

6) Program each of the math operations (a through j) shown in Figure 11-32 into the PLC, and prove the operation of each.

7) Design a program that will implement the following arithmetic operation:

- Using a move instruction, place the value of 16 in N7:1 and 48 in N7:2.
- Add the values together, and store the result in N7:3.

- Subtract the value in N7:3 from 650, and store the result in N7:4.
- Multiply the value in N7:4 by 15, and store the result in N7:5.
- Divide the value in N7:4 by 18, and store the result in F8:1.

Enter the program into the PLC, and prove its operation.

8) Create a program that will determine the average value of the accumulated value from four counters. Enter the program into the PLC and prove its operation.

9) A conveyor has 6-, 8-, and 12-packs of canned soda entering it. Each size of entering pack has an individual pack quantity counter. To know how many cans enter the conveyor, set up a program for multiplying and then adding to give a total can count. Enter the program into the PLC, and prove its operation.

10) Write a program that will implement the following arithmetic operation:

- Using a move instruction, place the value 30 in N7:1 and 25 in N7:2.
- Multiply the values together, and store the result in N7:3.
- Add the value 115 to the value stored in N7:3.
- Subtract the value 325 from the value stored in N7:3, and store the result in N7:4.
- Divide the value in N7:3 by 5, and store the result in F8:1.

Enter the program into the PLC, and prove its operation.

11a) Create a program that uses a file arithmetic logic (FAL) instruction to copy a table or file of data from N7:0–4 to N7:5–9. Add the two files together, and store the results at N7:10–14.

   b) Repeat part **a** using subtraction, multiplication, division, and square root expressions.

12) Two parts-conveyor lines, A and B, feed a main conveyor line M. A third conveyor line, R, removes rejected parts a short distance down from the main conveyor. Conveyors A, B, and R have parts counters connected to them. Construct a PLC program to obtain the total parts output of main conveyor M. Enter the program into the PLC, and prove its operation.

CHAPTER 12   Sequencer and Shift Register Instructions

TEST 12.1

Choose the letter that best completes the statement.

1. Which of the following would not be classified as a sequencer switch?   1._____

a) Rotary switch   c) Drum switch

b) Pressure switch   d) Stepper switch

2. Sequencer switches are used whenever   2.__D___

a) a counter function is required.

b) a timer function is required.

c) a time-delay function is required.

d) a repeatable operating pattern is required.

3. The information for each PLC sequencer step is entered into   3.__D___

a) the output module.   c) the programmer.

b) the input module.   d) a word file.

4. As the PLC sequencer advances through its steps, information is transferred from   4.__D___

a) the output module to the input module.

b) the input module to the output module.

c) the programmer to the processor.

d) the word file to the output word.

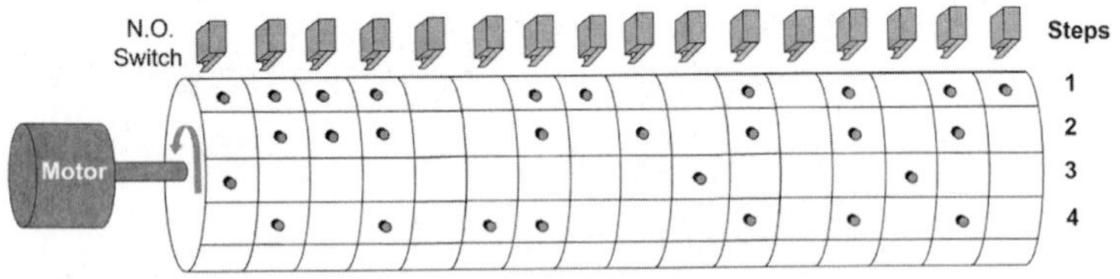

Figure 12-1 Sequencer data for question 5.

5. The equivalent sequencer data for step 2 of the sequencer shown in Figure 12-1 would be   5._____

a) 0000000000000000.   c) 0111001010101010.

b) 1111111111111111.   d) 1111 0000 0011011.

244

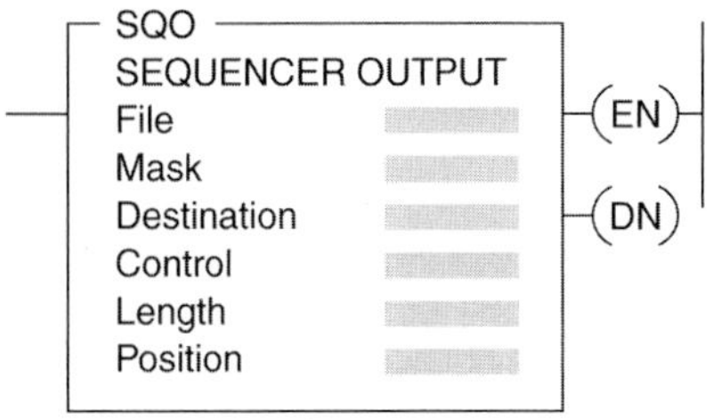

Figure 12-2 Sequencer Output instruction for question 6.

6-1. For the Sequencer Output (SQO) instruction of Figure 12-2, the ____ is the address of the output word to which the sequencer moves the data. 6-1. B

a) Control
b) Destination
c) Length
d) Position

6-2. The ____ is the address of the word used to selectively screen out data. 6-2. C

a) File
b) Length
c) Mask
d) Destination

6-3. The ____ is the starting address for the registers that contain the data to be transferred. 6-3. A

a) File
b) Control
c) Mask
d) Length

6-4. The ____ is the number of steps of the sequencer file. 6-4. D

a) Position
b) Control
c) Mask
d) Length

6-5. The ____ bit is set after the last word in the sequencer file is transferred. 6.5 D

a) Position
b) Control
c) EN
d) DN

Output word

	15	14	13	12	11	10	9	8	7	6	5	4	3	2	1	0
O:2	0	0	0	0	0	0	0	0	0	0	0	0	0	0	0	0

Sequencer file

	15	14	13	12	11	10	9	8	7	6	5	4	3	2	1	0	Positions
B3:0	0	0	0	0	0	0	0	0	0	0	0	0	0	0	0	0	Start
B3:1	1	0	0	1	1	0	0	1	0	0	1	0	0	0	0	1	Step1
B3:2	0	0	1	0	0	1	1	0	1	1	0	1	0	0	0	1	Step2
B3:3	0	1	0	1	0	1	0	0	0	0	0	0	1	1	0	0	Step3
B3:4	1	1	1	1	1	1	1	1	1	1	1	1	1	1	1	1	Step4

Figure 12-3 Sequencer Output instruction for question 7.

7-1. The sequencer instruction shown in Figure 12-3 reads the 16-bit data file words

a) one bit at a time.

b) one word at a time.

c) two bits at a time.

d) two words at a time.

7-1._____

7-2. As the sequencer advances through the steps, data are transferred

a) through the sequencer file from B3:0 to B3:4.

b) through the sequencer file from B3:4 to B3:0.

c) from the output word to the sequencer file.

d) from the sequencer file to the output word.

7-2._____

7-3 _____ outputs are to be controlled from one 16-point output module.

a) 6

b) 15

c) 16

d) 24

7-3._____

7-4. In step 1 _____ will be energized.

a) outputs O:2.0, O:2.5, O:2.8, O:2.11, O:2.12 and O:2.15

b) outputs O:2.2, O:2.3, O:2.10, O:2.12, and O:2.14

c) outputs O:2.0, O:2.4, O:2.6, O:2.7, O:2.9, O:2.10, and O:2.13

d) all outputs

7-4._A__

7-5. In step 2 _____ will be energized.

a) outputs O:2.0, O:2.5, O:2.8, O:2.11, O:2.12 and O:2.15

b) outputs O:2.2, O:2.3, O:2.10, O:2.12, and O:2.14

c) outputs O:2.0, O:2.4, O:2.6, O:2.7, O:2.9, O:2.10, and O:2.13

d) all outputs

7-5._C__

7-6. In step 3 _____ will be energized.

a) outputs O:2.0, O:2.5, O:2.8, O:2.11, O:2.12 and O:2.15

b) outputs O:2.2, O:2.3, O:2.10, O:2.12, and O:2.14

c) outputs O:2.0, O:2.4, O:2.6, O:2.7, O:2.9, O:2.10, and O:2.13

d) all outputs

7-6._B__

7-7. In step 4 _____ will be energized.

7-7._____

a) outputs O:2.0, O:2.5, O:2.8, O:2.11, O:2.12 and O:2.15

b) outputs O:2.2, O:2.3, O:2.10, O:2.12, and O:2.14

c) outputs O:2.0, O:2.4, O:2.6, O:2.7, O:2.9, O:2.10, and O:2.13

d) all outputs

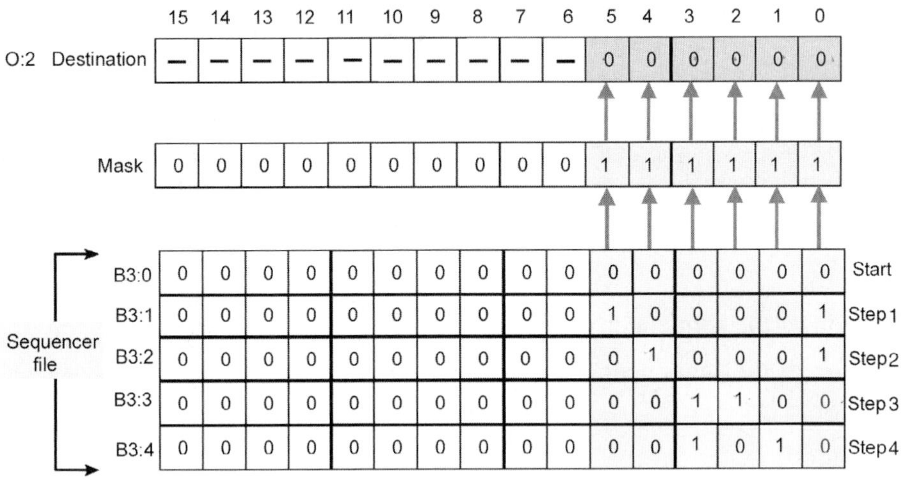

		15	14	13	12	11	10	9	8	7	6	5	4	3	2	1	0	
O:2	Destination	—	—	—	—	—	—	—	—	—	—	0	0	0	0	0	0	
	Mask	0	0	0	0	0	0	0	0	0	0	1	1	1	1	1	1	
	B3:0	0	0	0	0	0	0	0	0	0	0	0	0	0	0	0	0	Start
	B3:1	0	0	0	0	0	0	0	0	0	0	1	0	0	0	0	1	Step1
Sequencer file	B3:2	0	0	0	0	0	0	0	0	0	0	0	1	0	0	0	1	Step2
	B3:3	0	0	0	0	0	0	0	0	0	0	0	0	1	1	0	0	Step3
	B3:4	0	0	0	0	0	0	0	0	0	0	0	0	1	0	1	0	Step4

Figure 12-4 Sequencer Output instruction for question 8.

8-1. The sequencer instruction shown in Figure 12-4 controls _____ bits of output word O:2.

8-1. B

a) 2

c) 10

b) 6

d) 15

8-2. For each bit that the sequencer is to control, the corresponding bit of mask must be set to

8-2._____

a) 0.

c) positive.

b) 1.

d) negative.

8-3. All unmasked bits

8-3._____

a) can be used elsewhere in the program.

b) cannot be used elsewhere in the program.

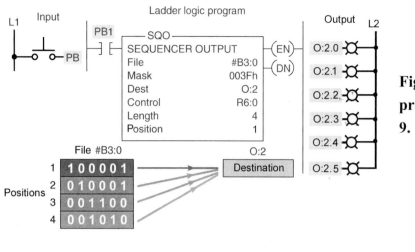

Figure 12-5 Sequencer program for question 9.

9-1. The sequencer program shown in Figure 12-5 advances to the next position on each

a) true to false transition of the sequencer rung.

b) false to true transition of the sequencer rung.

c) on power up of the PLC.

d) on power down of the PLC.

9-1._____

9-2. When the sequencer is at step 1, the binary information transferred into word O:2 would be

a) 010001.   c) 100001.

b) 001010.   d) 001010.

9-2._____

9-3. When the sequencer is at step ___, outputs O:2/2 and O:2/3 will be on and all the rest will be off.

a) 1   c) 3

b) 2   d) 4

9-3._C_

9-4. When the position parameter reaches 4, the ____ bit is set to 1.

a) Control   c) DN

b) EN   d) all of these

9-4._C_

10. Normally, sequencer instructions are retentive. This means that if you are at step 3 of a 10-step sequencer and power to your system is lost, when it is restored, the power sequencer will

a) return to step 3.   c) reset to step 1 automatically.

b) increment to step 4 automatically.   d) advance to step 10 automatically.

10._A_

11. A single sequencer instruction may have an upper limit on   11._____
a) the number of steps that can be programmed.
b) the number of external outputs that can be programmed.
c) the number of times the operating cycle can be actuated.
d) both a and b

12. When using a time-driven sequencer, the sequencer advances to the   12._____
next step
a) when the preset value equals the accumulated value.
b) when the preset value is less than the accumulated value.
c) for every true-to-false transition of the sequencer rung.
d) for every false-to-true transition of the sequencer rung.

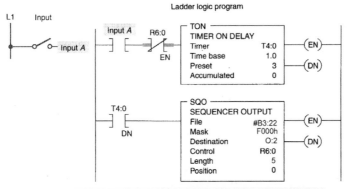

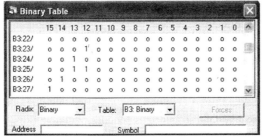

Figure 12-6 Sequencer program for question 13.

13-1. The sequencer program of Figure 12-6 is   13-1. _B_
a) motor-driven.   c) event-driven.
b) time-driven.   d) gear-driven.

13-2. The sequencer operates whenever   13-2.____
a) input A is false.   c) the PLC is in the run mode.
b) input A is true.   d) the PLC is in the program mode.

249

13-3. When the sequencer is functioning, the circuit increments   13-3.____
automatically through the ____ steps of the sequencer at ____ intervals.
a) five, 3-second   c) five, 3-minute
b) six, 3-second   d) six, 3-minute

13-4. When the sequencer is at position 2, which output(s) will be energized? 13-4.____
a) O:2/12   c) O:2/12 and O:2/13
b) O:2/13   d) O:2/14

13-5. A program modification is to be made to the sequencer that   13-5.____
requires outputs O:2/12, O:2/13, O:2/14, and O:2/15 to all be energized
at step 3. This would require
a) bits 12, 13, 14, and 15 of word B3:22 to be set to 1.
b) bits 13, 14, and 15 of word B3:23 to be set to 1.
c) bits 12, 14, and 15 of word B3:24 to be set to 1.
d) bits 14 and 15 of word B3:25 to be set to 1.

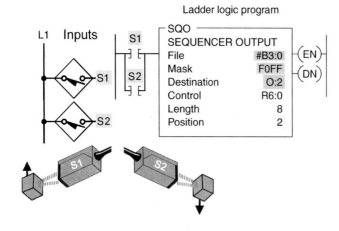

Ladder logic program

Figure 12-7 Sequencer program for question 14.

14-1. The sequencer program of Figure 12-7 is   14-1.____
a) motor-driven.   c) event-driven.
b) time-driven.   d) gear-driven.

14-2. The sequencer advances its position whenever   14-2.____
a) S1 makes a false to true transition
b) S2 makes a false to true transition
c) the PLC is in the run mode.
d) either a or b

14-3. For each position data are copied from file #B3:0 through the 14-3.____
_____ word to the destination.

a) mask

c) position

b) control

d) length

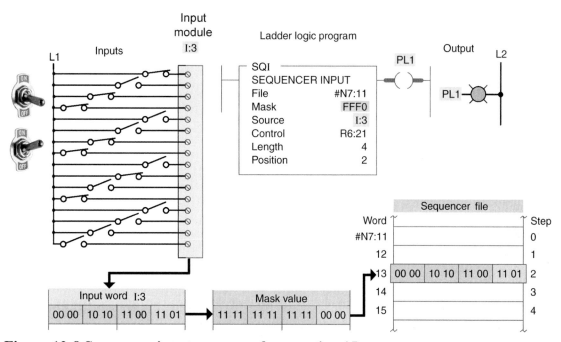

Figure 12-8 Sequencer input program for question 15.

15-1. The sequencer input instruction of Figure 12-8 is true whenever 15-1.____
the unmasked input data are _____ the data stored in the sequencer file.

a) matched to

c) less than

b) greater than

d) greater than or equal to

15-2. In this example the data at data at position _____ matches the unmasked 15-2.____
input data making the PL1 output _____.

a) 2, true

c) 4, true

b) 2, false

d) 4, false

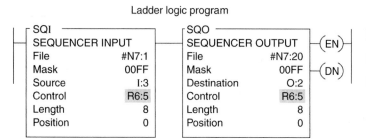

Figure 12-9 Sequencer input and output instructions for question 16.

16. When the SQI instruction is paired with the SQO instruction, as illustrated in Figure 12-9,

16._____

a) the same control address, length value, and position value are used for each instruction.

b) the sequencer input instruction is indexed by the sequencer output instruction.

c) it allows input and output sequences to function in unison.

d) all of these

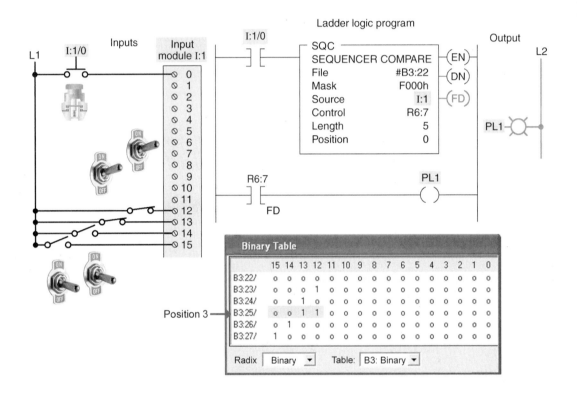

Figure 12-10 Sequencer compare instructions for question 17.

17-1. The sequencer compare instruction (SQC) of Figure 12-10 increments the position parameter whenever

17-1.____

a) the unmasked input data match the data stored in the sequencer file.

b) the unmasked input data do not match the data stored in the sequencer file.

c) input I:1/0 makes a false to true transition.

d) input I:1/0 makes a true to false transition.

17-2. The found bit (FD) is set true whenever the

17-2.____

a) instruction is true.

b) source pattern matches the sequencer file word.

c) source pattern does not match the sequencer file word.

d) PLC is powered on.

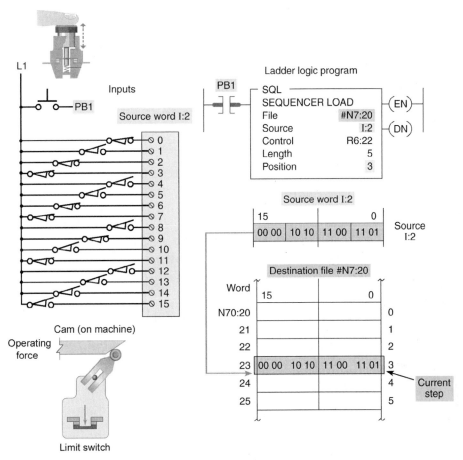

Figure 12-11 Sequencer load instructions for question 18.

18-1. The sequencer load instruction (SQL), shown in Figure 12-11, ____
data from the source to the sequence file.

18-1.____

a) adds

c) copies

b) subtracts

d) negates

18-2. The sequencer load instruction

18-2.____

a) does not function during the machine's normal operation.

b) replaces manual loading of data into the sequencer file.

c) does not use a mask.

d) all of these

19. A bit shift register shifts bits   19._____

a) serially from bit to bit.   c) randomly from bit to bit.

d) serially between words.   d) randomly between word.

20. A common application for a shift register would be   20._____

a) tracking parts.

b) controlling machine or process operations.

c) inventory control.

d) all of these

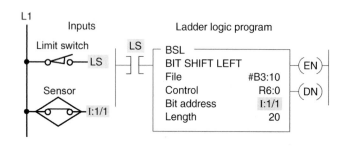

Figure 12-12 Bit shift left register for question 21.

	15	14	13	12	11	10	9	8	7	6	5	4	3	2	1	0
B3:10/	1	1	0	0	1	1	0	0	1	1	0	1	1	0	0	0
B3:11/	0	0	0	0	0	0	0	0	0	0	0	0	1	1	1	0

B3: Table - Before limit switch clock pulse

21-1. For the bit shift left (BSL) register shown in Figure 12-12, file   21-1. _B_
length is given in

a) words.   c) steps.

b) bits.   d) files.

21-2. Momentary actuation of ____ causes the BSL instruction to execute.   21-2.____

a) the limit switch

b) the sensor

c) either the limit switch or sensor

d) both the limit switch and the sensor

21-3. The data block contains   21-3. _C_

a) 20 words.   c) 20 bits.

b) 32 words.   d) 32 bits.

21-4. When the BSL rung goes from false to true, the data block is shifted   21-4.____

a) one bit position to a lower bit number.

b) one bit position to a higher bit number.

c) two bit positions to a lower bit number.

d) two bit positions to a higher bit number.

21-5. Each time the BSL executes, the last bit is   21-5.____

a) shifted out of the array.

b) reset to 0.

c) set to 1.

d) shifted to the start of the array.

21-6. All bits in the unused portion of the last word of the file   21-6.____

a) can be used elsewhere in the program.

b) can only be used elsewhere in the program if masked.

c) can only be used elsewhere in the program if not masked.

d) should not be used elsewhere in the program.

22. A bit shift register operates synchronously in that   22.__D__

a) information is shifted one bit at a time within a word or words.

b) for every bit shifted in, one is shifted out.

c) the data entered must be shifted the length of the register before they are available to be shifted out.

d) all of these

23. In a word shift register, the data are shifted out ___bit(s) at a time.   23._____

a) 1   c) 4

b) 2   d) 16

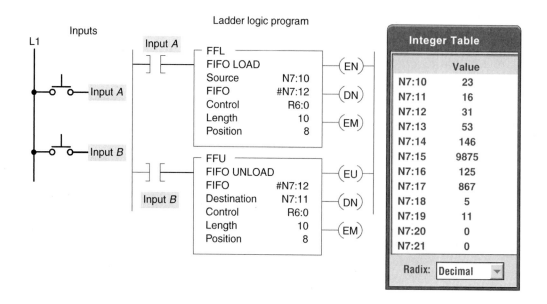

Figure 12-13 Word shift register program for question 24.

24-1. For the word shift register program shown in Figure 12-13 the _____ instruction loads logic words into FIFO stack.

24-1._____

a) FFL

c) EM

b) FFU

d) EN

24-2. The address of the stack is:

24-2._____

a) R6:0.

c) #N7:12.

b) N7:10.

d) N7:11.

24-3. Data enter the FIFO file form _____ on a false to true transition of input A.

24-3._____

a) R6:0

c) #N7:12

b) N7:10

d) N7:11

24-4. A false to true transition of input B causes all data in the FIFO file to

24-4. B

a) shift one position toward the ending address of the file.

b) shift one position toward the starting address of the file.

c) set all data in the file to 1.

d) reset all data in the file to 0.

256

TEST 12.2

Place the answers to the following questions in the answer column at the right.

1. Mechanical sequencer switches are often referred to as (a) ____ , (b) ____ ,
(c) ____ , or (d) ____ switches.

1a._____
1b._____
1c._____
1d._____

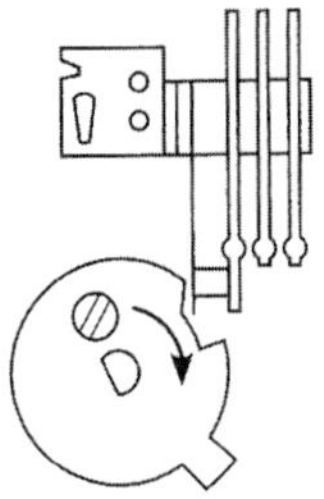

Figure 12-14 Mechanical sequencer switch for question 2.

2-1. With the mechanical sequencer switch shown in Figure 12-14, contacts
interact with the cam so that, for different degrees of rotation of the cam,
different contacts (a) ____ and (b) ____ .

2-1a.____
2-1b.____

2-2. An electric ____ is used to rotate the sequencer cam.

2-2._____

3. Sequencer switches can be used for processes that require a repeatable
operating pattern. (True or False)

3._____

4. To program a sequencer, binary information is entered into a series of
consecutive memory ____ .

4._____

5. As a programmed sequencer advances through its steps, binary
information is transferred from the sequencer file to the ____ word.

5._____

6. When a sequencer operates on an entire output word, all outputs
associated with the word are required to be controlled by the sequencer.
(True or False)

6._____

7. Bits of an output word not used by the sequencer can be used elsewhere
in your program. (True or False)

7._____

8. The ____ word selectively screens out data from the sequencer word file to the output word.

8._____

9. Sequencers, like other PLC instructions, are programmed exactly the same for all PLC models. (True or False)

9._____

10. Due to the way in which a sequencer instruction operates, all output points must be on a single output module. (True or False)

10._____

11. Sequencer instructions simplify your ladder program. (True or False)

11._____

12. Sequencer instructions are usually nonretentive. (True or False)

12._____

13. There is usually no limit to the number of external outputs and steps that can be operated on by a single sequencer instruction. (True or False)

13._____

14. Many sequencer instructions reset the sequencer automatically to step 1 on completion of the last sequence step. (True or False)

14._____

15. A(n) ____-driven sequencer operates in a manner similar to a mechanical drum switch that increments automatically after a preset time period.

15._____

16. A(n) ____-driven sequencer advances to the next step by an external pulsed input.

16._____

17. The hexadecimal equivalent of the binary number 0011 1111 is ____.

17._____

18. With time-driven sequencers, each step functions in a manner similar to timer instructions because it involves a(n) (a) ____ time value and a(n) programmed (b) ____ time value.

18a._____
18b._____

19. The ____ parameter is the address of the sequencer file.

19._____

20. Before a sequencer starts its sequence, we need a starting point where the sequencer is in a neutral position. The start position is all zeroes, representing our neutral position; thus all outputs will be ____.

20._____

21. The sequencer _____ parameter is where the status bits, length, and instantaneous position are stored.

21._____

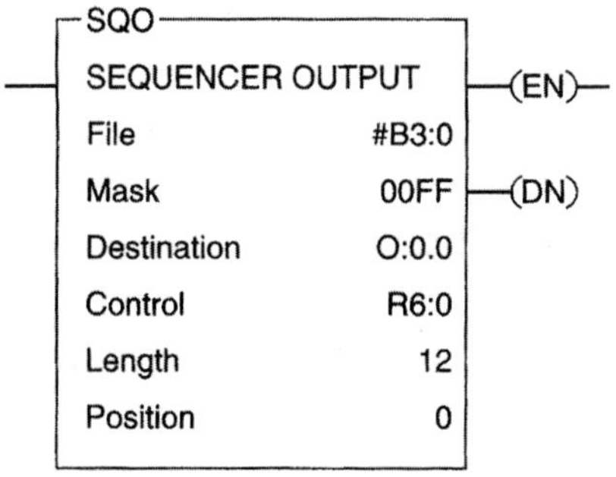

Figure 12-15 Sequencer output instruction for question 22.

22. The sequencer output instruction of Figure 12-15 will send the data in file (a) _____ out to the (b) _____ output connections through the (c) _____ value. The length of the file is specified as 12 (d) _____.

22a._____
22b._____
22c._____
22d._____

23. State which sequencer instruction (SQO, SQC, or SQL) would be used if you want to (a) capture reference conditions by manually stepping the machine through its operating sequence; (b) control sequential machine operations by transferring 16-bit data to output image addresses; (c) monitor machine operating conditions for diagnostic purposes by comparing 16-bit image data with data in a reference file.

23a._____
23b._____
23c._____

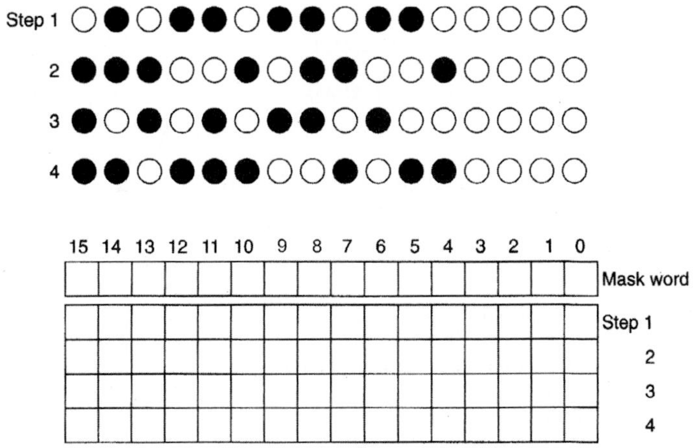

Figure 12-16 Sequencer output file for question 24.

24. Complete the information for the mask word and sequencer file in Figure 12-16 so that the sequencer will operate the lamps as shown (dark circle indicates lamp is on).

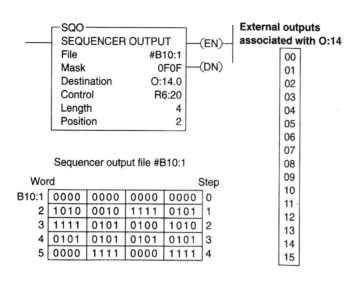

Sequencer output file #B10:1

Word					Step
B10:1	0000	0000	0000	0000	0
2	1010	0010	1111	0101	1
3	1111	0101	0100	1010	2
4	0101	0101	0101	0101	3
5	0000	1111	0000	1111	4

Figure 12-17 Sequencer data for question 25.

25. For the SQO instruction condition of Figure 12-17, which outputs will be on?

25._____

26. Shift registers are often used for _____ parts on a production line.

26._____

27. In general the two types of shift instructions are bit shift (a) _____ and bit shift (b) _____.

27a._____
27b._____

28. With a bit shift register, the status data (1 or 0) is shifted automatically through the register from one bit address to the next. (True or False)

28._____

29. Shift registers cannot be used to control processes where parts are shifted continually from one position to the next. (True or False)

29._____

30. You can program a shift register instruction to shift only 1s through the register. (True or False)

30._____

31. If you wanted to produce an external output when a certain bit in a shift register is on, you would program a rung with an examine for on instruction corresponding to the _____ address.

31._____

32. When you program a shift register instruction, you can shift data only to the left. (True or False)

32._____

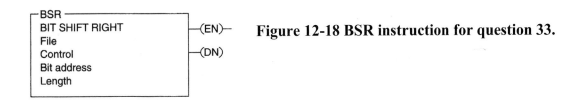

Figure 12-18 BSR instruction for question 33.

33. For the BSR instruction of Figure 12-18, which parameter tells the processor the

a) instruction's address?

b) number of bits in the bit array?

c) source bit address?

d) location of the bit array?

33a._____

33b._____

33c._____

33d._____

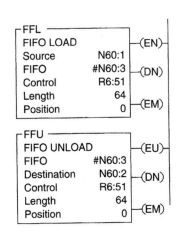

Figure 12-19 FFL-FFU instruction pair for question 34.

34. For the FFL-FFU instruction pair of Figure 12-19, which parameter tells the processor

a) the location of the exit word?

b) the location of the next in word?

c) to start at the FIFO file address?

d) the location of the stack?

e) the instruction's address?

f) the maximum number of words you can load?

34a._____

34b._____

34c._____

34d._____

34e._____

34f._____

Programming Assignments

This section requires you to simulate how PLC sequencer and shift register functions operate. The instructions used are intended to be generic in nature and, as such, will require some conversion for the particular PLC model you are using. The use of a prewired PLC input/output control panel is recommended to simulate the operation of these circuits.

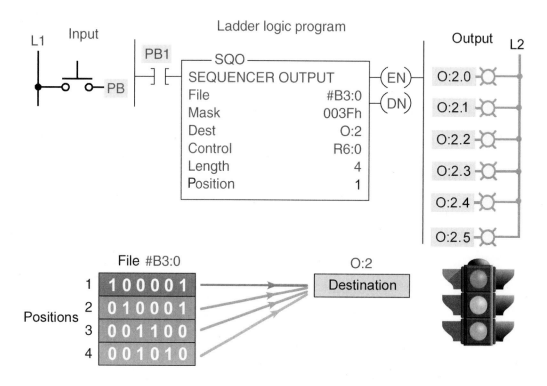

Figure 12-20 Event-driven sequencer for assignment 1.

1) The event-driven traffic light sequencer program of Figure 12-20 is described in the text. Prepare an I/O connection diagram and ladder logic program that will simulate its operation. Use addresses that apply to your PLC installation. Enter the program into the PLC, and verify its operation by using the pushbutton to step manually the sequencer steps.

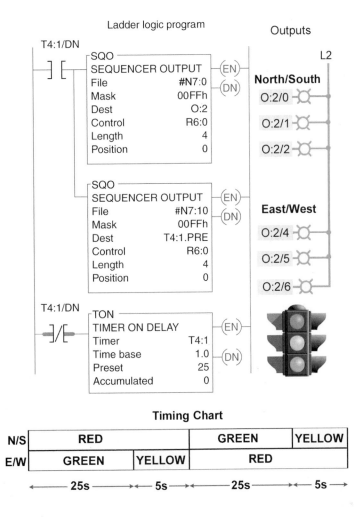

Figure 12-21 Time-driven sequencer for assignment 2.

2) The time-driven traffic light sequencer program of Figure 12-21 is described in the text. Prepare an I/O connection diagram and ladder logic program that will simulate its operation. Use addresses that apply to your PLC installation. Enter the program into the PLC, and verify its operation.

3a) Design a PLC sequencer program that provides the following:

- The sequencer must step in 5-s intervals.
- Output No. 1—on all the time the machine is cycling.
- Output No. 2—on except for steps 3 and 5.
- Output No. 3—on only in step 3
- Output No. 4—on in steps 2 and 4.
- Output No. 5—on in steps 2, 3, and 4.

- Output No. 6—on in steps 1 and 5.

Prepare a sequencer file, I/O connection diagram, and ladder logic program for the circuit. Enter the program into the PLC, and prove its operation.

b) Modify the program to provide the following additional control features:

- All outputs must stay off and the sequencer must not operate until a start button is pressed.
- Once the start button is pressed, the sequencer completes one complete cycle and then stops automatically.
- Pushing a stop button resets and stops the sequencer.

Prepare a sequencer file, I/O connection diagram, and ladder logic program for the circuit. Enter the program into the PLC, and prove its operation.

4) Traffic flow on a one-way street is to be controlled by means of a pedestrian pushbutton so that the green traffic light and the Don't Walk pedestrian light are to be normally on at all times when the pedestrian pushbutton is not actuated; and when the pedestrian pushbutton is actuated, the sequencer is started and controls the outputs as follows:

- The green traffic light immediately switches off, and the amber traffic light switches on to begin to stop the traffic flow—the Don't Walk pedestrian light remains on. Outputs remain in this state for 5 s.
- The amber traffic light switches off, and the red traffic light switches on—the Don't Walk pedestrian light remains on. Outputs remain in this state for 5 s to ensure that traffic has stopped before pedestrians begin to cross.
- The Don't Walk pedestrian light switches off, and the Walk pedestrian light switches on—the red traffic light remains on. Outputs remain in this state for 15 s, allowing pedestrians safe passage across the street.
- The Walk pedestrian light switches off, and the Don't Walk pedestrian light switches on—the red traffic light remains on. Outputs remain in this state for 5 s to ensure that pedestrians are not still crossing the street when the traffic light changes from red to green.
- The green traffic light switches on, and the red traffic light switches off—the Don't Walk pedestrian light remains on. Outputs remain in this state for 30 s to ensure a

minimum amount of automobile traffic flow time, even if the walk pushbutton is frequently actuated.

- The sequencer stops, and the green traffic light and Don't Walk pedestrian light remain on until the pedestrian pushbutton is pressed to start the cycle again.

Prepare a sequencer file, I/O connection diagram, and ladder logic program that can be used to simulate this traffic control system. Enter the program into the PLC, and prove its operation.

Event	Water input	Soap release	Hot wax	Air blower
LS1	1	0	0	0
LS2	1	1	0	0
LS3	1	0	0	0
LS4	0	0	1	0
LS5	0	0	0	1
LS6	0	0	0	0

Figure 12-22 Car wash sequencer chart for assignment 5.

5) Design a PLC event-driven sequencer output program for the automatic car wash sequencer chart of Figure 12-22. Enter the program into the PLC, and prove its operation.

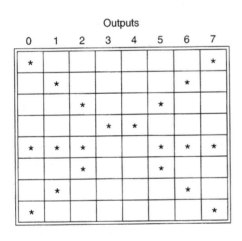

Figure 12-23 Matrix chart for assignment 6.

6a) Design a PLC sequencer program to turn on outputs according to the matrix chart shown in Figure 12-23. Use a switch to step through the table. Enter the program into the PLC, and prove its operation.

b) Modify the program to operate continuously by using a recycling timer's done bit to trigger a step in the sequence. Enter the modified program into the PLC, and prove its operation.

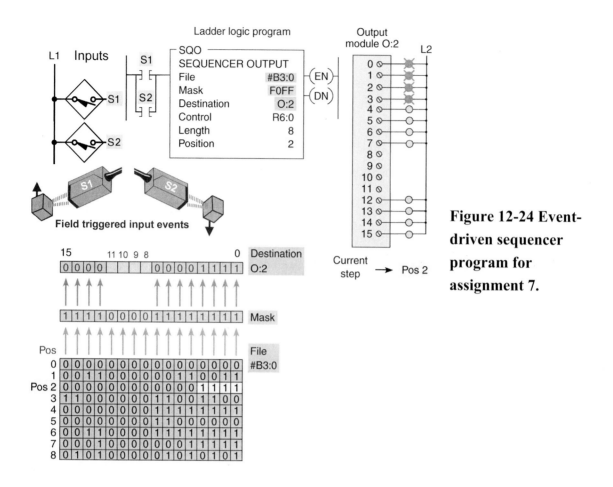

Figure 12-24 Event-driven sequencer program for assignment 7.

7) The event-driven sequencer program of Figure 12-24 is described in the text. Prepare an I/O connection diagram and ladder logic program that will simulate its operation. Use addresses that apply to your PLC installation. Enter the program into the PLC, and verify its operation.

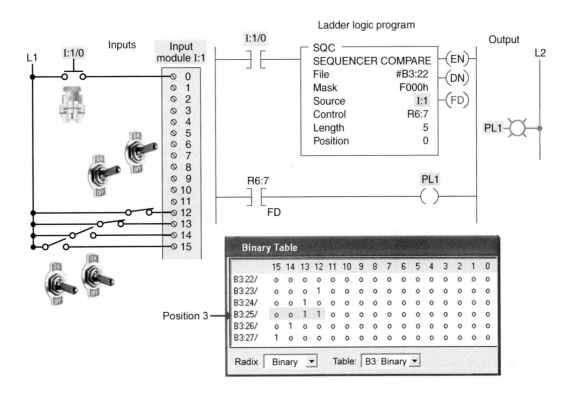

Figure 12-25 Sequencer input (SQC) program for assignment 8.

8) The sequencer compare (SQC) program of Figure 12-25 is described in the text. Prepare an I/O connection diagram and ladder logic program that will simulate its operation. Use addresses that apply to your PLC installation. Enter the program into the PLC, and verify its operation.

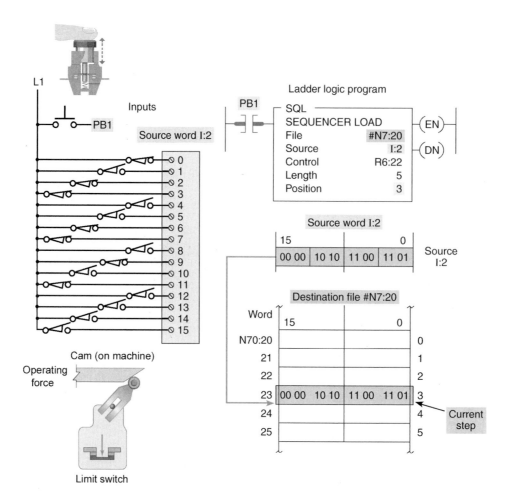

Figure 12-26 Sequencer load (SQL) program for assignment 9.

9) The sequencer load (SQL) program of Figure 12-26 is described in the text. Prepare an I/O connection diagram and ladder logic program that will simulate its operation. Use addresses that apply to your PLC installation. Enter the program into the PLC, and verify its operation.

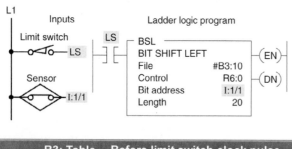

Figure 12-27 Bit shift left (BSL) program for assignment 10.

B3: Table - Before limit switch clock pulse															
	15	14	13	12	11	10	9	8	7	6	5	4	3	2	1	0
B3:10/	1	1	0	0	1	1	0	0	1	1	0	1	1	0	0	0
B3:11/	0	0	0	0	0	0	0	0	0	0	0	0	1	1	1	0

10) The bit shift left (BSL) program of Figure 12-27 is described in the text. Prepare an I/O connection diagram and ladder logic program that will simulate its operation. Use addresses that apply to your PLC installation. Enter the program into the PLC, and verify its operation.

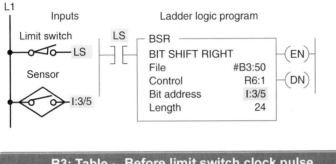

Figure 12-28 Bit shift right (BSR) program for assignment 11.

	15	14	13	12	11	10	9	8	7	6	5	4	3	2	1	0
B3: Table - Before limit switch clock pulse																
B3:50/	1	0	1	1	0	0	0	1	1	0	0	1	0	1	1	0
B3:51/	0	0	0	0	0	0	0	0	1	0	1	1	0	0	1	1

11) The bit shift right (BSR) program of Figure 12-28 is described in the text. Prepare an I/O connection diagram and ladder logic program that will simulate its operation. Use addresses that apply to your PLC installation. Enter the program into the PLC, and verify its operation.

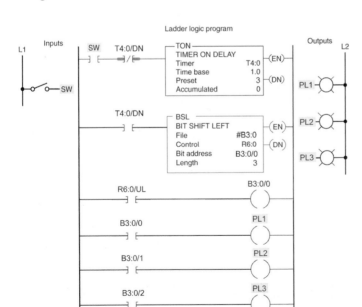

Figure 12-29 Wraparound bit shift left program for assignment 12.

12) The wraparound bit shift left (BSL) program of Figure 12-29 is described in the text. Prepare an I/O connection diagram and ladder logic program that will simulate its operation. Use addresses that apply to your PLC installation. Enter the program into the PLC, and verify its operation.

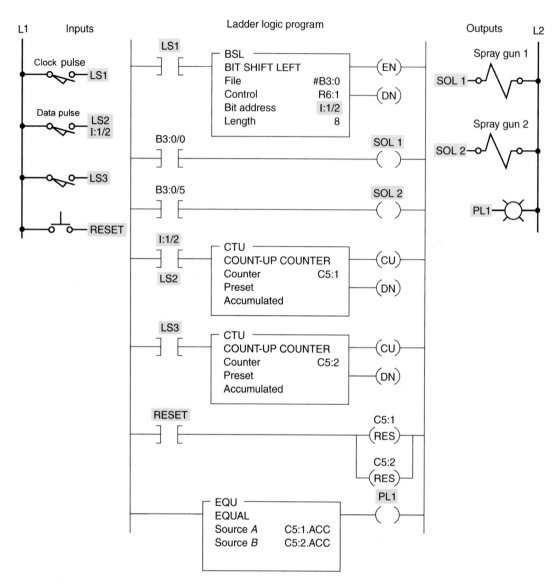

Figure 12-30 Spray-painting program for assignment 13.

13) The spray-painting program of Figure 12-30 is described in the text. Prepare an I/O connection diagram and ladder logic program that will simulate its operation. Use addresses that apply to your PLC installation. Enter the program into the PLC, and verify its operation.

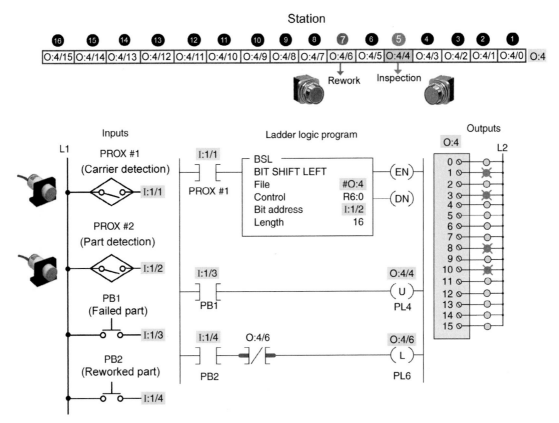

Figure 12-31 Carrier tracking program for assignment 14.

14) The carrier tracking program of Figure 12-31 is described in the text. Prepare an I/O connection diagram and ladder logic program that will simulate its operation. Use addresses that apply to your PLC installation. Enter the program into the PLC, and verify its operation.

15) Construct a program that will keep track of the presence of parts on a 23-station conveyor line *as follows:*

- If a part is placed on the line, then a limit switch connected to input A address will close.
- The conveyor will be indexed by pressing a pushbutton connected to input B.
- An indicator light connected to output C will turn on when a part comes off the line.

Enter the program into the PLC, and prove its operation.

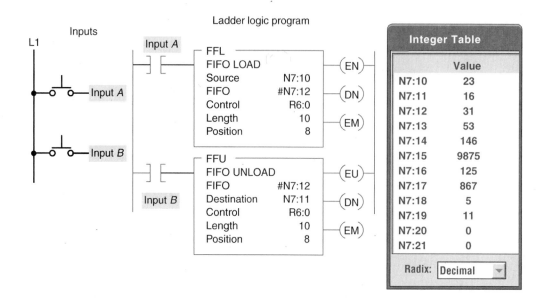

Figure 12-32 FIFO word shift register program for assignment 16.

16) The FIFO word shift register program of Figure 12-32 is described in the text. Prepare an I/O connection diagram and ladder logic program that will simulate its operation. Use addresses that apply to your PLC installation. Enter the program into the PLC. and verify its operation.

CHAPTER 13   PLC Installation Practices, Editing, and Troubleshooting

TEST 13.1

Choose the letter that best completes the statement.

1. PLCs are placed within an enclosure to provide protection against   1. ___D___
a) atmospheric conditions.   c) moisture.
b) conductive dust.   d) all of these

2. For most PLC installations, a NEMA ____ enclosure is recommended.   2._____
a) 2   c) 8
b) 4   d) 12

3. Typically, PLC systems installed inside an enclosure can withstand a   3._____
maximum of
a) 60°C outside the enclosure.   c) 60°C inside the enclosure.
b) 50°C outside the enclosure.   d) 50°C inside the enclosure.

4. Malfunctions due to electrical noise interference usually cause   4.___A___
a) temporary occurrences of operating errors.
b) permanent occurrences of operating errors.
c) temporary loss of memory.
d) permanent loss of memory.

5. A good location for a PLC enclosure is close to   5._____
a) the machine or process.   c) high-frequency welders.
b) large AC motors.   d) annealing furnaces.

6. When routing power and I/O signal wiring to and within a PLC   6.___D___
enclosure, you should
a) use the shortest possible wire runs for I/O signals.
b) never run signal and power wiring in the same conduit.
c) run low level signal conductors as shielded twisted pair.
d) all of these

7. Electrical noise can be coupled into a PLC system   7. _D_
a) by an electrostatic field.   c) by a fiber optic system.
b) through electromagnetic induction.   d) both a and b

8. Which of the following would not normally be located within the   8._____
PLC enclosure?
a) I/O modules   c) Master control relay
b) Limit switch   d) Isolation transformer

9. Under no circumstances should   9._____
a) fiber optic and power wiring be run in the same conduit.
b) fiber optic and signal wiring be run in the same conduit.
c) signal wiring and power wiring be run in the same conduit.
d) all of the above

10. Certain input field devices may have a small leakage current when   10._____
they are
a) in the ON state.   c) examined for an ON condition.
b) in the OFF state.   d) examined for an OFF condition.

11. A leakage problem can occur when connecting an output module to a   11._____
high-impedance load. This problem can be corrected by connecting
a) a bleeder resistor in series with the load.
b) a bleeder resistor in parallel with the load.
c) an NO contact in series with the load.
d) an NC contact in parallel with the load.

12. I/O leakage problems usually occur with devices that use   12._____
a) solid-state switching circuits.   c) noise-suppression circuits.
b) hard contacts.   d) voltage-suppression circuits.

13. The authoritative source on grounding requirements for a PLC   13._____
installation is the
a) plant electrician.   c) equipment manufacturer.
b) plant engineer.   d) National Electrical Code.

14. In addition to being an important safety measure, proper grounding of   14._____
a PLC system can
a) lower installation costs.
b) increase the power efficiency of the system.
c) limit the effects of EMI.
d) assist in the operation of the MCR.

15. Proper grounding procedures for a PLC installation specify that   15.__D__
a) all enclosures, CPU and I/O chassis, and power supplies be connected
to a single low-impedance ground.
b) paint or nonconductive materials should be scraped away to provide good
ground connections.
c) all ground connections should be made using star washers.
d) all of these

16. Which of the following load devices is most likely to require some   16.__C__
form of noise or voltage suppression?
a) Lamp   c) Solenoid
b) Heater   d) LED display

17. Excessive line voltage variations to a PLC installation can be corrected   17.__A__
by installing a
a) constant voltage transformer.   c) step-up transformer.
b) step-down transformer.   d) current transformer.

18. A high voltage spike is generated whenever current to   18.__A__
a) an inductive load is turned off.   c) a resistive load is turned off.
b) an inductive load is turned on.   d) a resistive load is turned on.

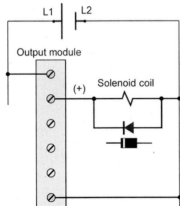

Figure 13-1 Suppression circuit for question 19.

19. The diode used in the suppression circuit of Figure 13-1   19._____

a) is connected in forward-bias to suppress DC loads.

b) is connected in reverse-bias to suppress DC loads.

c) is connected in forward-bias to suppress AC loads.

b) is connected in reverse-bias to suppress AC loads.

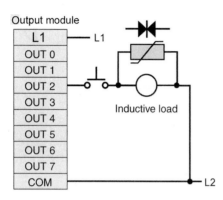

Figure 13-2 Suppression circuit for question 20.

20. The suppression circuit of Figure 13-2   20._____

a) can be used to suppress DC loads.   c) uses a metal oxide surge suppressor.

b) can be used to suppress AC loads.   b) all of these

21. Editing a PLC program normally refers to   21._____

a) replacement of an existing program.

b) monitoring of an existing program.

c) making changes to an existing program.

d) all of these

22. Preparing a PLC control process for start-up is called   22._____

a) commissioning.   c) validating.

b) troubleshooting.   d) installing.

23. In which of the following program modes are modifications   23._____
executed immediately on entry of the instruction?

a) Continuous test mode   c) Off-line program mode

b) Single scan test mode   d) On-line program mode

24. The data monitor feature of a PLC may allow you to   24._____

a) change the radix or data format.   c) view data within an instruction.

b) store data for an instruction prior to use.   d) all of these

25. Which function of a PLC would be used to search each instance of   25._____
a particular address?
a) Change radix   c) Cross reference
b) Forcing   d) Contact histogram

26. Which function of a PLC would be used to view the transition history   26._____
(ON or OFF states) of a bit(s)?
a) Change radix   c) Cross reference
b) Forcing   d) Contact histogram

27. Which of the following would not normally be included as part of   27._____
a routine preventive maintenance program?
a) Inspection of I/O field devices   c) Checking connections to I/O modules
b) Monitoring the program   d) Cleaning of the PLC enclosure

28. If an output module fuse blows repeatedly, a probable cause may be   28.__D__
a) the module's output current is being exceeded.
b) the output device may be shorted.
c) the output field wiring may be shorted.
d) any of these

29. Which of the following is not normally included in the array of status   29.__B__
indicators found on the processor module?
a) Memory OK   c) Battery OK
b) Wiring OK   d) Power supply OK

30. A watchdog timer is used to monitor   30.__A__
a) how long it takes the CPU to complete a scan.
b) battery voltage.
c) memory circuits.
d) all of these

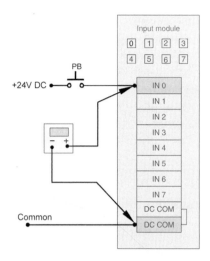

Figure 13-3 Input module circuit for question 31.

31. With reference to the input module circuit of Figure 13-3, a voltmeter reading of 24 VDC would indicate

31._____

a) normal operation.

c) a possible fault in the field wiring.

b) a possible shorted PB.

d) either b or c

	Input device condition	Input module status indicator	Monitor display status indicator	
			False	True
(a)	Closed — ON 24V DC Input	ON	⊣⊢	⊣/⊢
(b)	Closed — ON 0V DC Input	OFF	⊣⊢	⊣/⊢
(c)	Open — OFF 0V DC Input	OFF	⊣⊢	⊣/⊢
(d)	Open — OFF 24V DC Input	ON	True ⊣⊢	False ⊣/⊢

Figure 13-4 Guide for question 32.

32. With reference to the troubleshooting guide of Figure 13-4, which condition indicates no fault?

32._____

Input device condition	Input module status indicator	Monitor display status indicator	
		True	False
Open — OFF 24V DC Input	ON	⊣⊢	⊣/⊢

Figure 13-5 Fault condition for question 33.

33. With reference to the condition illustrated in Figure 13-5, what is the most likely fault?   33._____
a) Short circuit in the field device or wiring
b) Input module fault
c) Processor module fault
d) Open in the field device or wiring

34. The input wiring or device is suspected as being the source of a PLC problem. If this circuit is not at fault, then the status indicator on the input module should be illuminated when the input device is   34._____
a) closed and programmed for an XIC condition.
b) open and programmed for an XIO condition.
c) closed, regardless of how it is programmed.
d) open, regardless of how it is programmed.

35. Assume an open in the field wiring is suspected between the output module and output load device. This condition would be confirmed if   35.__A__
a) full output voltage was measured at the module and 0 voltage at the load.
b) full output voltage was measured at the load and 0 voltage at the module.
c) the output status indicator on the module is on and the load is not operating.
d) the output status indicator on the module is off and the load is not operating.

	Output device condition	Output module status indicator	Monitor display status indicator
(a)	De-energized — off	On	True —( )—
(b)	De-energized — off	Off	True —( )—
(c)	Energized — on	On	True —( )—
(d)	De-energized — off	Off	False —( )—

Figure 13-6 Guide for question 36.

36. Which of the choices of the troubleshooting guide shown in Figure 13-6   36._____
indicates a problem with the wiring to the output device or the output device itself?

37. The majority of PLC system faults are caused by   37.__*D*__
a) faulty power supplies.   c) software failure.
b) malfunctioning microprocessors.   d) field wiring and devices.

38. A single output device has failed while the remainder of the PLC   38._____
system is functioning normally. The indicator light on the output module
indicates that a signal is sent to the output point where the device is
connected. You would now
a) trace the circuit back through the logic to locate the inputs.
b) use a programming terminal to call up the rung that controls the output to
see if the output coil is on.
c) check the point where the output device's field wiring is connected to the
output rack.
d) check the input modules for short-circuit conditions.

39. The force instruction: 39._____
a) will force data table bits on only.
b) will force data table bits off only.
c) if used indiscriminately, could cause haphazard machine operation.
d) can be used only to force inputs.

40. The first step in troubleshooting is to   40.__A__
a) identify or describe the faulty operation.
b) test the process field devices.
c) test the wiring.
d) test the I/O modules.

41. The I/O module input and output status lights   41._____
a) are found only on analog modules.
b) indicate the status of the inputs and outputs.
c) indicate whether the process field devices are faulty.
d) are not used in the troubleshooting process.

42. Which instruction is used to change the amount of logic scanned to progressively debug a program?

42._____

a) Suspend
c) Force
b) Temporary end
d) Timed interrupt

43. Program addressing conflicts can be caused by

43._____

a) using the same address for two outputs.
c) faulty connections.
b) faulty wiring.
d) scan time period.

44. With Allen-Bradley controllers ____ software is required to develop and edit programs.

44._____

a) RSLinx
c) Microsoft word
b) RSLogix
d) LogixPro

45. With Allen-Bradley controllers ____ software is required to monitor, download, and upload programs.

45._____

a) RSLinx
c) Microsoft word
b) RSLogix
d) LogixPro

46. PLC communications protocol is a standardized method for

46. _D_

a) programming PLCs.
b) connection of input devices.
c) connection of output devices.
d) transmitting data between different devices.

TEST 13.2

Place the answers to the following questions in the answer column at the right.

1. PLCs are generally placed within an enclosure. (True or False)   1._____

2. An enclosure is used to shield the controller from electrical (a) ____   2a._____
and airborne (b)____.   2b._____

3. Most PLC installations require additional cooling provisions, not included   3._____
in the original installation. (True or False)

4. A hardwired electromechanical master control relay is not normally   4._____
included as part of the wiring for a PLC system. (True or False)

5. PLC malfunctions due to electrical noise usually produce temporary   5._____
occurrences of operating errors. (True or False)

6. Electromagnetic interference (EMI) may enter a PLC system through   6a._____
either (a) ____ or (b) ____.   6b._____

7. Common potential noise generating devices include noninductive   7._____
resistive loads. (True or False)

8. A fiber optic wired control system is most susceptible electrical noise.   8._____
(True or False)

9. Running signal wiring and power wiring in the same conduit helps   9._____
cut down on electrical noise. (True or False)

10. Most solid-state switches will conduct a small amount of leakage   10._____
current in the ____ state.

11. Leakage current can falsely activate a PLC input. (True or False)   11._____

12. A ____ resistor can be connected to drain off unwanted leakage current.   12._____

13. A properly installed grounding system will provide a ____-impedance path to earth ground.

13._____

14. When a grounding connection must be made to the metal enclosure of a PLC system, any paint coating must remain intact. (True or False)

14._____

15. Ground loop circuits cause problems by adding or subtracting current or voltage from input signal sources. (True or False)

15._____

16. A ground loop circuit can develop when each device's ground is tied to a different earth potential. (True or False)

16._____

17. Where line voltage variations to the PLC are excessive, a(n) ____ voltage transformer can be used to maintain a steady voltage.

17._____

18. When current in to a DC ____ load is interrupted or turned off, a very high voltage spike is generated.

18._____

19. A(n) ____ can be connected in reverse-bias across a DC solenoid to suppress voltage spikes.

19._____

20. Generally, output modules designed to drive inductive loads include suppression networks as part of the module circuit. (True or False)

20._____

21. A metal oxide varistor (MOV) can be used to suppress AC or DC voltages. (True or False)

21._____

22. Using PLC editing functions, instructions and rungs can be added, deleted, or modified. (True or False)

22._____

23. Preparing a PLC control process for initial start-up is called ____.

23._____

24. The offline programming mode permits the user to change the program during the machine operation. (True or False)

24._____

25. When a PLC is placed in the continuous test mode, the processor operates the user program without ____ any outputs.

25._____

26. PLC data _____ functions allow you to monitor and/or modify specified program variables.

26._____

27. The contact histogram function is used to locate each instance of a selected address. (True or False)

27._____

28. The cross reference function allows you to view the transition history of a data table value. (True or False)

28._____

29. All field I/O devices should be inspected periodically to ensure that they are properly adjusted as an important part of a PLC's _____ maintenance program.

29._____

30. The processor of a PLC monitors its battery voltage level. (True or False)

30._____

31. To make sure that equipment does not operate while PLC maintenance is conducted, (a) _____ and (b) _____ devices must be in use.

31a._____
31b._____

32. The first step in the troubleshooting of a PLC system is to identify the (a) _____ and its (b) _____.

32a._____
32b._____

33. One of the diagnostic checks carried out by the processor is the proper operation of all I/O devices. (True or False)

33._____

34. The watchdog timer is a separate timing circuit that must be set and reset by the _____.

34._____

35. If the scan time is too short, a watchdog error will be declared. (True or False)

35._____

36. Usually, each I/O device has at least two status indicators. One of these indicators is on the I/O (a) _____, and the other is provided by the programming device (b) _____.

36a._____
36b._____

37. The status indicator on an input module will normally be illuminated if the input device is off and examined for an off condition. (True or False)

37._____

38. The programming device monitor normally indicates a true instruction if the addressed input device is off and examined for an OFF condition. (True or False)

38._____

39. PLC output modules may incorporate fuse or electronic protection for output circuitry. (True or False)

39._____

40. It is rare for sensors and actuators connected to the I/O of the process to fail. (True or False)

40._____

41. The suppression device is wired in ____with the load device.

41._____

42. The temporary end instruction is used when you want to change the amount of logic being scanned. (True or False)

42._____

43. The suspend instruction is used to trap and identify specific conditions. (True or False)

43._____

44. An addressing conflict is caused by the ____ address being used for two or more coil instructions in the same program.

44._____

45. The PLC will only accept one program at a time. (True or False)

45._____

46. Communications ____ is a standardized method for establishing communications between different PLC devices.

46._____

Programming Assignments

This section requires you to perform simulated PLC editing and troubleshooting functions.

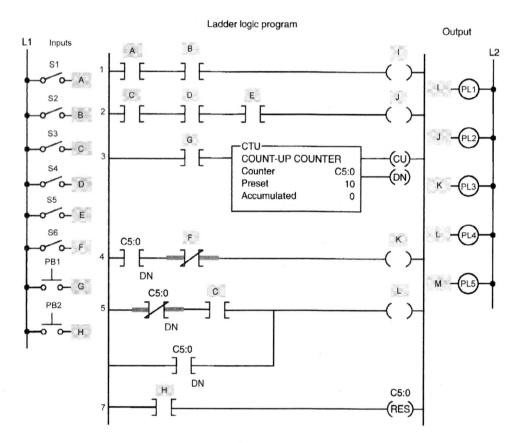

Figure 13-7 Editing and data control program for assignment 1.

1) Construct a simulated program for the program editing and data control exercise shown in Figure 13-7. Use field devices and addresses available for use with your PLC trainer. Have the program checked by your instructor after each editing change. Enter the original program into the PLC, and prove its operation.

- Enter an additional rung following rung No. 4. This new rung is to examine the status bit of output 1 for an OFF condition and energize output M when the rung condition is true.
- Remove the instruction from rung No. 2 that examines address E for an XIC condition.

- Place the XIC instruction (E) that was removed from rung No. 2 into rung No. 6. Insert this XIC instruction in parallel with the existing XIC instruction H.
- While in the run mode, change the preset value of the counter from 10 to 25.
- Remove rung No. 5 of the original program.
- While in the run mode, force output K to an ON condition.
- While in the run mode, force output J to an OFF condition.

2) Enter the original program of assignment 1 into the PLC, and complete the troubleshooting exercises that follow. Have your instructor simulate each type of fault. Demonstrate your ability to identify the source of each problem systematically.

a) Defective input module

b) Defective output module

c) Blown fuse in an output module

d) Shorted input device

e) Open input device

f) Open output device

g) Open in field wiring to an input device

h) Open in field wiring to an output device

CHAPTER 14 Process Control, Network Systems and SCADA

TEST 14.1

Choose the letter that best completes the statement.

1. A continuous process is   1._____
a) one that never shuts down.
b) used only for simple tasks.
c) one in which raw materials enter one end of the system and the finished
product comes out the other end.
d) used only with computers.

2. Assume two ingredients are added together, processed, and then   2._____
stored. This would be an example of a(n)
a) batch process.   c) individual product-producing process.
b) continuous process.   d) discrete product-producing process.

3. A distributive control system (DCS):   3._____
a) permits the distribution of the processing task among several controllers.
b) always utilizes a single large computer.
c) will stop the whole process if one control element fails.
d) is the least flexible type of control system.

4. Components of a control system may include   4._____
a) sensors.   c) human-machine interface (HMI).
b) actuators.   d) all of these

5. HMIs allow operators to _____ the application.   5._____
a) control   c) diagnose
b) monitor   d) all of these

6. Which of the following devices could be classified as a sensor?　　　6._____
a) Thermistor　　　　　　　　　　　c) Solenoid
b) Relay　　　　　　　　　　　　　　d) All of these

7. Which of the following devices could be classified as an actuator?　　7._____
a) Control valve　　　　　　　　　　c) Servo motor
b) Electric brake　　　　　　　　　　d) All of these

8. Generally, compared to an open-loop system, a closed loop is　　　8._____
a) more accurate.　　　　　　　　　　c) more expensive.
b) more complex.　　　　　　　　　　d) all of these

9. A closed-loop control system measures the _____ output of the　　　9._____
process and compares it to the _____ output.
a) actual, desired　　　　　　　　　　c) operating, nonoperating
b) no-load, full-load　　　　　　　　　d) final, initial

10. The set point for a control system refers to　　　　　　　　　10._____
a) the input that determines the operating point for the process.
b) a process variable that is monitored continually.
c) a process error that is uncontrolled.
d) all of these

11. A closed-loop control system　　　　　　　　　　　　　　11._____
a) requires less power to operate.
b) does not require a feedback signal from the process.
c) uses a feedback signal from the process.
d) requires more power to operate.

12. The error signal in a closed-loop control system is　　　　　　12._____
a) always a positive value.
b) always a negative value.
c) the difference between the set point and feedback signal.
d) the sum of the set point and feedback signal.

13. Which of these controller types provides the fastest response to a   13._____
system error?
a) PID   c) Proportional plus integral
b) On/off   d) Proportional plus derivative

14. With an on/off controller   14._____
a) the output is either completely on or completely off.
b) a positive deviation of the process variable from its set point causes the
controller to shut the control element off.
c) a negative deviation of the process variable from its set point causes the
controller to turn the control element on.
d) all of these

15. Time-proportioning control refers to   15._____
a) linear movement of the final control element.
b) varying the ratio of ON time to OFF of the final control element.
c) the integral action of a controller.
d) the derivative action of a controller.

16. A proportional controller   16._____
a) is designed to eliminate the cycling associated with on/off control.
b) allows the final control element to take intermediate positions between on
and off.
c) permits analog control of the final control element.
d) all of these

17. The integral action responds to   17._____
a) the size and time duration of the error signal.
b) the speed at which the error signal is changing.
c) proportional bandwidth.
d) proportional gain.

18. The derivative action responds to   18._____
a) the size and time duration of the error signal.
b) the speed at which the error signal is changing.
c) proportional bandwidth.
d) proportional gain.

19. A PID controller   19._____

a) is tuned using a signal generator.

b) is factory-tuned for optimum performance.

c) must be custom-tuned to each process.

d) both a and b

20. Each motor of a PLC motion control system is referred to as   20._____

a) an axis of motion.   c) stepper motor.

b) a synchronous motor.   d) a control component.

21. The function of the servo drive as part of a PLC motion control   22._____
system is to

a) provide power to the servo motors.

b) translate signals from the motion module into motor drive commands.

c) monitor the servo motor's position and velocity.

d) all of these

22. Each axis of an industrial robot arm is controlled by   22._____

a) an open-loop servo motor system.

b) a closed-loop servo motor system.

c) an on/off controller.

d) a PID controller.

23. PLC system data communications is accomplished using   23._____

a) network links.

b) point-to-point serial communications links.

c) transformer links.

d) both a and b

24. Open communications networks   24._____

a) are based on standards developed through industry associations.

b) do not require that you to buy all components from a single supplier.

c) do not use a proprietary protocol.

d) all of these

25. The fundamental job of a local area network (LAN) is to provide _____ between devices.

    a) communication   c) isolation

    b) connections   d) protection

25._____

26. The transmission medium used in data communications is

    a) coaxial cable.   c) fiber optics.

    b) twisted pair.   d) all of these

26._____

27. Each device on an industrial network is known as a

    a) load.   c) node.

    b) control.   d) repeater.

27._____

28. Network _____ is refers to the physical layout of devices on a network.

    a) topology   c) reliability

    b) functionality   d) all of these

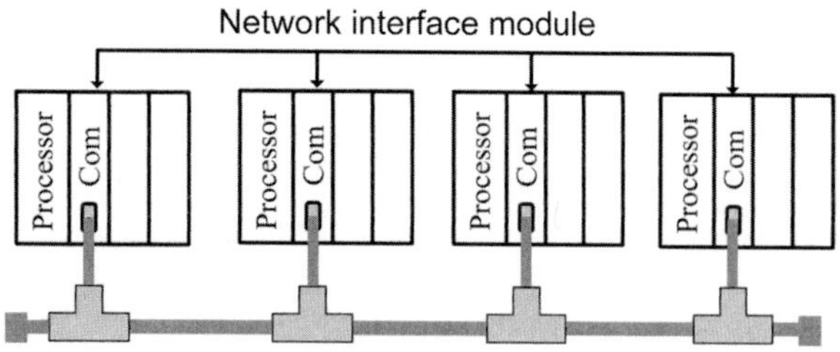

28._____

Figure 14-1 Network connection for question 29.

29. The type of network connection topology shown in Figure 14-1 is

    a) bus.   c) tree.

    b) star.   d) ring.

29._____

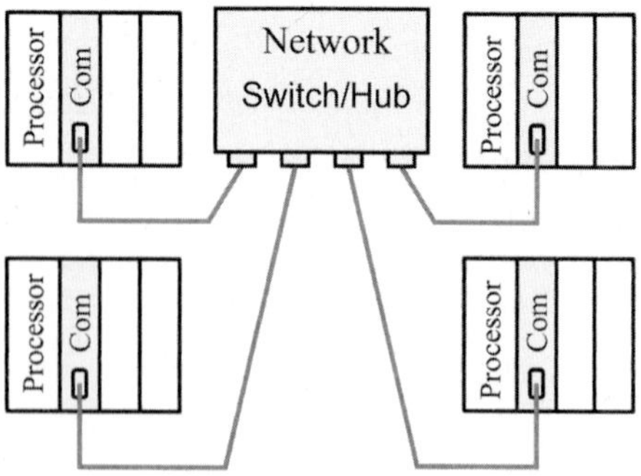

Figure 14-2 Network connection for question 30.

30. The type of network connection topology shown in Figure 14-2 is   30._____
a) bus.   c) tree.
b) star.   d) ring.

31. Network ____ defines how data are arranged and coded for transmission   31._____
on a network.
a) devices   c) protocol
b) medium   d) functions

32. Communication between different PLC architectures and protocols is   32._____
made possible by the use of
a) rectifiers.   c) gateways.
b) repeaters.   d) hubs.

33. In a token passing network access control scheme, a node can   33._____
transmit data on the network
a) at all times.   c) only at the end of a scan cycle.
b) only when it has possession of a token.   d) only at the start of a scan cycle.

34. In a collision detection network access control scheme, a node transmits   34._____
data on the network
a) at all times.
b) when other nodes are sending messages on the network.
c) at preset timed intervals.
d) if there are no other messages on the network.

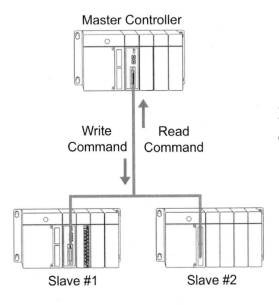

Figure 14-3 Master-slave network for question 35.

35. The network access control scheme used in the master-slave protocol   35._____
of Figure 14-3 is known as
a) polling.   c) token passing.
b) collision detection.   d) analog detection.

36. A peer-to-peer PLC network   36._____
a) uses the token passing access control scheme.
b) has no master PLC.
c) has each device identified by an address.
d) all of these

37. The two methods of transmitting PLC digital data are   37._____
a) AC and DC.   c) input and output.
b) serial and parallel.   d) negative and positive

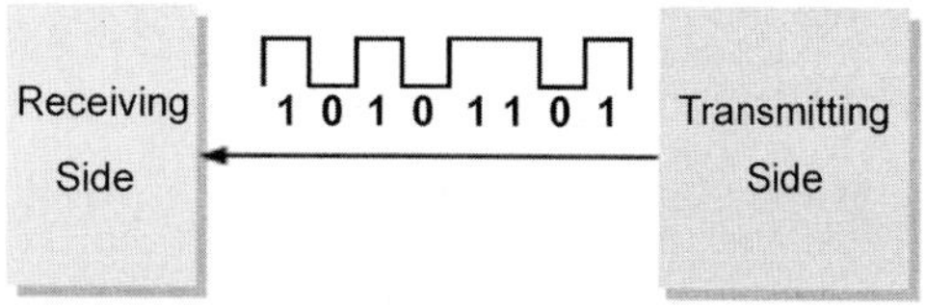

Figure 14-4 Data transmission for question 38.

38. Figure 14-4 illustrates an example of ____ data transmission.   38._____
a) DC   c) output
b) serial   d) positive

39. Which communication system allows communications simultaneously   39._____
in both directions?
a) Direct   c) Full-duplex
b) Indirect   d) Half-duplex

40. The Allen-Bradley data highway network   40._____
a) is a proprietary communications network.
b) uses peer-to-peer communications.
c) is implemented using token passing access.
d) all of these

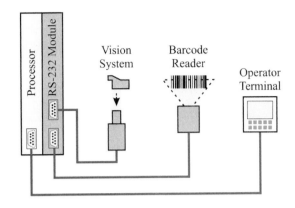

Figure 14-5 Communication connections for question 41.

41. Figure 14-5 illustrates as implementation of _____ type communication connections.

41._____

a) serial

c) analog

b) parallel

d) digital

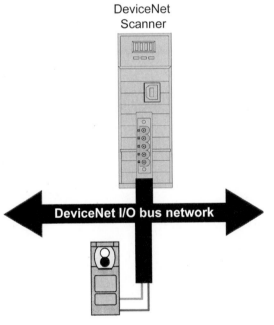

Figure 14-6 DeviceNet network for question 42.

42-1. The DeviceNet network shown in Figure 14-6 is used with

42-1.____

a) sensors.

c) valves.

b) switches.

d) all of these

42-2. This network

42-2.____

a) exchanges information with the field device.

b) supports field devices from various manufacturers.

c) supplies power to the field device.

d) all of these

42-3. The DeviceNet scanner is used instead of   42-3._____

a) I/O modules.   c) the chassis power supply module.

b) the processor module.   d) all of these

43. The ControlNet protocol   43._____

a) is an open network.

b) provides a high-speed link between controller and I/O devices.

c) is highly deterministic and repeatable.

d) all of these

44. The EtherNet/IP protocol   44._____

a) is a proprietary network.

b) will not operate with either DeviceNet or ControlNet protocol.

c) is based on the Control and Information protocol.

d) all of these

45. The Modbus protocol   45._____

a) transmits information over serial lines between devices.

b) uses the master-slave communication technique.

c) is an open protocol.

d) all of these

46. A Fieldbus communication system   46._____

a) can be implemented using daisy-chain topology.

b) is a proprietary system.

c) cannot serve as a network for field devices.

d) all of these

47. SCADA is an acronym that stands for   47._____

a) security control and data acquisition.

b) supervisory control and data acquisition.

c) security control and digital acquisition.

d) supervisory control and analog acquisition.

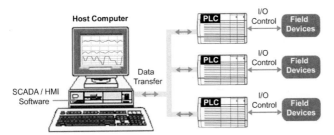

Figure 14-7 SCADA system for question 48.

48. For the SCADA system of Figure 14-7, the host computer   48._____

a) monitors the process.

b) sends commands to the PLCs.

c) stores data automatically.

d) all of these

TEST 14.2

Place the answers to the following questions in the answer column at the right.

1. A continuous process involves the flow of product material from
one section of the process to another. (True or False)

1._____

2. In a(n) _____ process, a set amount of product is received and then some
operation is performed on the product.

2._____

3. _____ control is used when several machines are controlled by
one controller.

3._____

4. _____ control involves two or more computers communicating with each
other to accomplish the complete control task.

4._____

5. One disadvantage of centralized control is that if the main controller
fails, the whole process is stopped. (True or False)

5._____

6. Distributive control systems (DCS) use one controller for all the
processing tasks. (True or False)

6._____

7. Distributive control systems are network-based. (True or False)

7._____

8. Actuators convert physical information into electrical signals.
(True or False)

8._____

9. A pushbutton switch could be classified as a type of
human-machine interface. (True or False)

9._____

10. A sensor could be classified as a type of controller. (True or False)

10._____

11. Signal _____ involves converting input and output signs into
a usable form.

11._____

12. _____ values of graphic HMI terminals display information on process
variables over a period of time.

12._____

298

13. HMI graphic terminal software is used to create and animate objects related to the process on the screen. (True or False)

13._____

14. Control systems can be classified as (a) _____ loop or (b) _____ loop.

14a._____
14b._____

15. A _____-loop system is one in which the output of a process affects the input control signal.

15._____

16. Sensors convert physical information into _____ signals.

16._____

17. In an open-loop control system, the controller receives no information concerning the status of the process. (True or False)

17._____

18. Closed-loop control contains a feedback element. (True or False)

18._____

19. On/off control eliminates hunting. (True or False)

19._____

20. With on/off control, the measured variable will _____ around the set point.

20._____

21. The _____ of a controller is the range above and below the set point that will not produce a change in the control action.

21._____

22. Deadband is used in controllers to prevent repeated activation-deactivation cycles. (True or False)

22._____

23. Proportional controllers are designed to eliminate the cycling associated with on/off control. (True or False)

23._____

24. On/off control permits analog control of the final control element. (True or False)

24._____

25. Time-proportioning control varies the ratio of (a) _____ time to (b) _____ time.

25a._____
25b._____

26. Proportioning action occurs within a proportional _____ around the set-point temperature.

26._____

27. The operation of a proportional controller leads to process deviation known as _____.

27._____

28. Integral action eliminates steady-state error. (True or False)

28._____

29. Derivative action responds to the _____ at which the error signal is changing.

29._____

30. A PID controller produces an output that depends on the (a) _____, (b) _____, and (c) _____of the system error signal.

30a._____
30b._____
30c._____

31. The _____ input determines the desired operating point for a process.

31._____

32. A P1D controller must be factory-tuned to each process being controlled. (True or False)

32._____

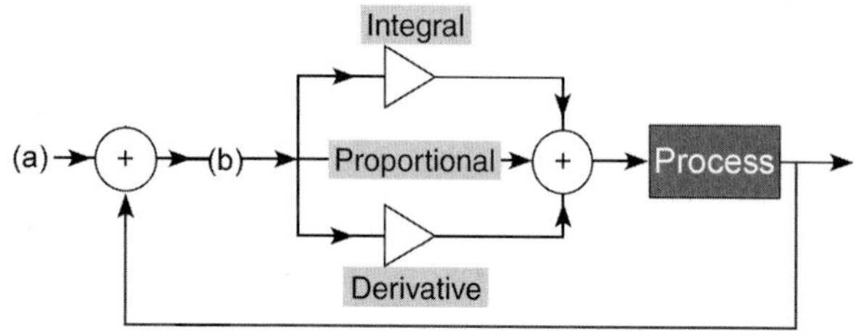

Figure 14-8 PID loop for question 33.

33. For the PID control loop of Figure 14-8, identify the signal (a) _____ and signal (b) _____.

33a._____
33b._____

34. A PID controller reduces the system error to zero faster than any other type of controller. (True or False)

34._____

35. A PID loop is normally tested by making an abrupt change to the set point and observing the controller's response rate. (True or False)

35._____

36. A fuzzy logic PID controller changes the amount of output signal in a mathematically specified way. (True or False)

36._____

37. PLCs can be fitted with I/O modules that produce PID control, or may have sufficient mathematical functions that allow PID control to be carried out. (True or False)

37._____

38. PLCs can be used for both (a) ____ and (b) ____ motion control applications.

38a._____
38b._____

39. Basic control components of a PLC motion control system are (a) ____, (b) ____, (c) ____, and (d) ____.

39a._____
39b._____
39c._____
39d._____

40. A robot arm is basically a series of mechanical links driven by ____ motors.

40._____

41. The two general types of communications links found in PLC systems are (a) ____ and (b) ____.

41a._____
41d._____

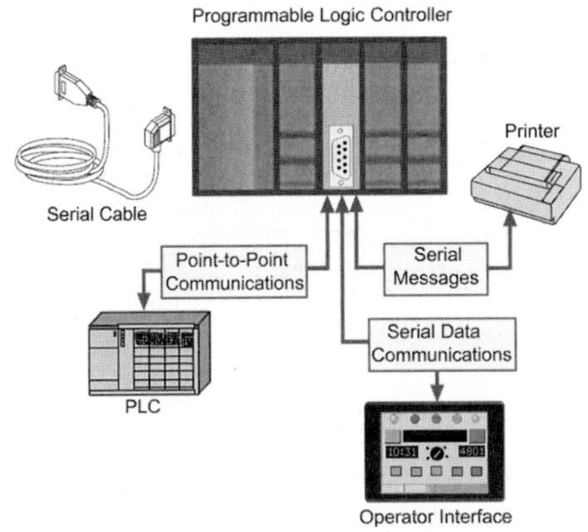

Figure 14-9 Communication links for question 42.

42. Figure 14-9 illustrates examples of point-to-point ____ communication links.

42._____

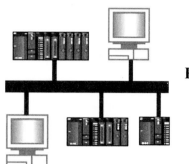

Figure 14-10 Communication link for question 43.

43. Figure 14-10 illustrates an example of a local area ____
communication link.

43._____

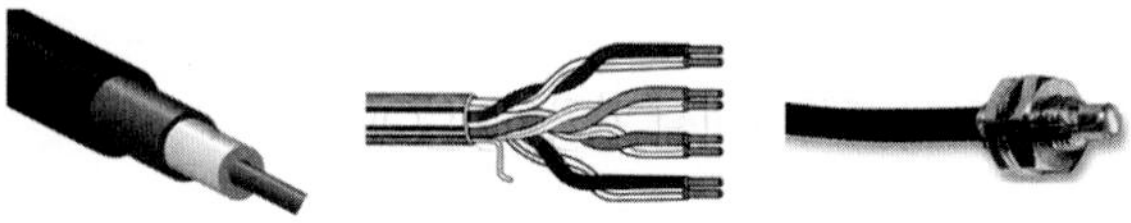

Figure 14-11 Communication link for question 44.

(a)   (b)   (c)

44. For the transmission media of Figure 14-11, identify each type shown in
(a) ____, (b) ____, and (c) ____.

44a._____
44b._____
44c._____

45. PLC networks are not able to communicate via wireless radio wave
systems. (True or False)

45._____

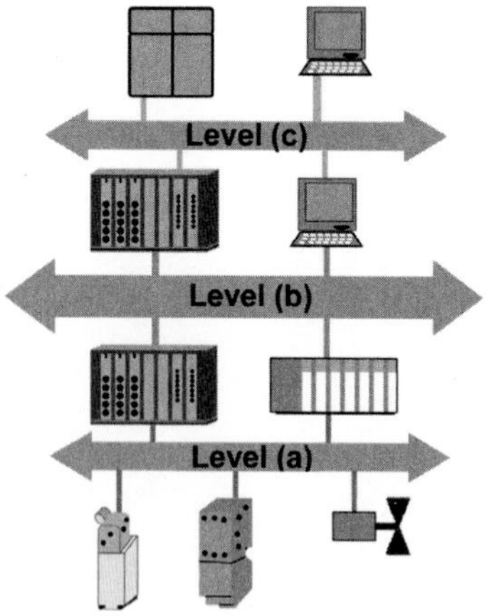

Figure 14-12 Industrial network for question 46.

Level (c)

Level (b)

Level (a)

46. For the industrial network of Figure 14-12, identify the level of
functionality for (a) ____, (b) ____, and (c) ____.

46a._____
46b._____
46c._____

47. A network _____ is a device that amplifies a signal to its original strength.   47._____

48. Each device connected on a network is known as a _____.   48._____

49. Network _____ refers to the physical layout of devices on a network.   49._____

50. A network switch or hub is required for network _____ topology.   50._____

51. Bus topology is a network configuration in which all stations are   51._____
connected in _____.

52. _____ bus networks interface with devices such as pushbuttons.   52._____

53. Network _____ defines how data are arranged and coded for transmission   53._____
on a network.

54. Gateways make communication possible between different protocols.   54._____
(True or False)

55. The _____ method refers to the manner in which a PLC accesses a bus   55._____
network to transmit information.

56. In a token passing based network, a node can transmit data on the   56._____
network at all times. (True or False)

57. Ethernet networks use a(n) _____ detection based access control scheme.   57._____

58. In master-slave polling protocol network, direct communications among   58._____
slaves is possible. (True or False)

59. Peer-to-peer networks use the token passing media access method.   59._____
(True or False)

60. In _____ transmission data are transferred one bit at a time.   60._____

61. _____-duplex transmission allows the transmission of data in both   61._____
directions simultaneously.

62. The Allen-Bradley data highway network is an open communications network. (True or False)

62._____

63. Serial transmission is recommended for distances of over 50 feet. (True or False)

63._____

64. DeviceNet is a proprietary high-speed device level network. (True or False)

64._____

65. The field devices connected to a DeviceNet network contain intelligence in the form of a microprocessor. (True or False)

65._____

66. ControlNet is an open high-speed network that is highly deterministic and repeatable. (True or False)

66._____

67. EtherNet/IP is an open communications network based on the same protocol that is used with DeviceNet and ControlNet. (True or False)

67._____

68. ____ refers to the data rate of a network expressed in terms of bits per second.

68._____

69. Both Modbus and Fieldbus are serial communication protocols. (True or False)

69._____

70. A SCADA system usually refers to a system that coordinates but does not ____ processes in real time.

70._____

Programming Assignment

1) Connect two PLC slave stations to one PLC master station, and assign nodes to each station. Create three ladder logic programs, and download them to the appropriate PLC. Verify the operation of the network.

CHAPTER 15   ControlLogix Controllers

Part 1   Memory and Project Organization

TEST 1

Choose the letter that best completes the statement.

1. The memory organization of a ControlLogix (CLX) controller
a) has fixed areas of memory for specific types of data.
b) has fixed areas of memory for inputs and outputs.
c) uses a flexible memory structure with no areas allocated for specific types of data.
d) both a and b

1._____

2. Configuration of a modular CLX system involves
a) establishing a communications link between the controller and the process.
b) identifying the type of processor used.
c) identifying the type of I/O modules used.
d) all of these

2._____

3. The _____ is the network browser interface that provides a single window
to view all configured network drivers.
a) RSLogix programming software
b) RSLinx
c) RSWho
d) all of these

3._____

4. A project
a) is not required in a CLX application.
b) contains all the information related to the CLX application.
c) is contained in the task in a CLX application.
d) is executed based on an event.

4._____

5. What are the major components of a project?   5._____
a) Main routine, subroutine, and fault routine
b) Continuous tasks and periodic tasks
c) Tasks, programs, and routines
d) all of these

6. The RSLogix 5000 Controller Organizer   6._____
a) is a tree-style presentation of the entire project.
b) simplifies the navigation and the overall view of the whole project.
c) presents all the information about the programs, data, and I/O configuration
of the current project.
d) all of these

7. Each folder of the controller organizer tree is expanded by   7._____
a) clicking on the + sign in front of the folder.
b) clicking on the − sign in front of the folder.
c) placing the controller in the Run mode.
d) placing the controller in the Program mode.

8. A task is a   8._____
a) scheduling mechanism for executing programs.
b) file that stores the logic for a controller.
c) file that stores the data for a controller.
d) file that stores the configuration for a controller.

9. A continuous task   9._____
a) executes nonstop.   c) has the lowest priority.
b) is always interrupted by a periodic task.   d) all of these

10. An event task is triggered   10._____
a) automatically.   c) by an event that failed to happen.
b) by an event that happened.   d) either b or c

11. A routine is   11._____
a) a set of logic elements for a specific programming language.
b) where the programmer writes the executable code for the project.
c) specified as ladder logic, sequential function chart, function block, or structure text.
d) all of these

12. Which routine is configured to execute first when the program runs?   12._____
a) Fault   c) Main
b) Subroutine   d) Start

13. Which routine is configured to be called by another routine?   13._____
a) Fault   c) Main
b) Subroutine   d) Start

14. ControlLogix controllers use _____ to refer to memory locations.   14._____
a) numbers   c) routines
b) tags   d) predefined data tables

15. Which type of tag can only be accessed by routines within a   15._____
specific program?
a) Base tag   c) Program scope tag
b) Alias tag   d) Controller scope tag

16. A(n) _____ tag is one whose value is received from another controller. 16._____
a) Base   c) Consumed
b) Alias   d) Produced

17. A(n) _____ tag defines a memory location where data are stored.   17._____
a) Base   c) Consumed
b) Alias   d) Produced

18. A(n) _____ tag is one that the controller makes available for use by 18._____
one or more other controllers.
a) Base   c) Consumed
b) Alias   d) Produced

19. A(n) _____ tag refers to a memory location defined by another existing tag. 19._____

a) Base   c) Consumed

b) Alias   d) Produced

20. Logix controllers are based on _____-bit operation.   20._____

a) 8   c) 32

b) 18   d) 64

21. A SINT base tag uses _____ bits of memory.   21._____

a) 8   c) 32

b) 18   d) 64

22. A structure type tag   22._____

a) is a grouping of different data types.   c) is made up of members.

b) functions as a single unit.   d) all of these

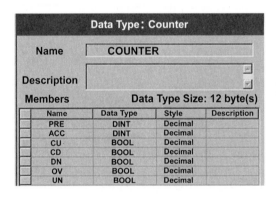

Figure 15-1 Data structure for question 23.

23. The type of data structure shown in Figure 15-1 would be classified as   23._____

a _____ type.

a) predefined   c) user-defined

b) module-defined   d) strictly defined

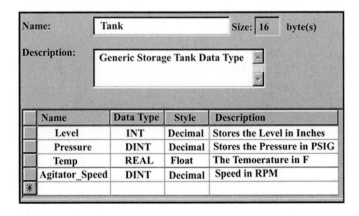

Figure 15-2 Data structure for question 24.

24. The type of data structure shown in Figure 15-2 would be classified as   24._____

a ____ type.

a) predefined   c) user-defined

b) module-defined   d) strictly defined

Figure 15-3 Data structure for question 25.

25. The type of data structure shown in Figure 15-3 would be classified as   25._____

a ____ type.

a) predefined   c) user-defined

b) module-defined   d) strictly defined

26. Tags are created in a CLX controller   26._____

a) using the tag editor before your program is entered.

b) by entering tag names as you program.

c) by using question marks in place of tag names and assigning names later.

d) any of these

27. Which of the following pieces of information is not optional when   27._____

defining a tag?

a) Tag name   c) Tag description

b) Tag type   d) Data type

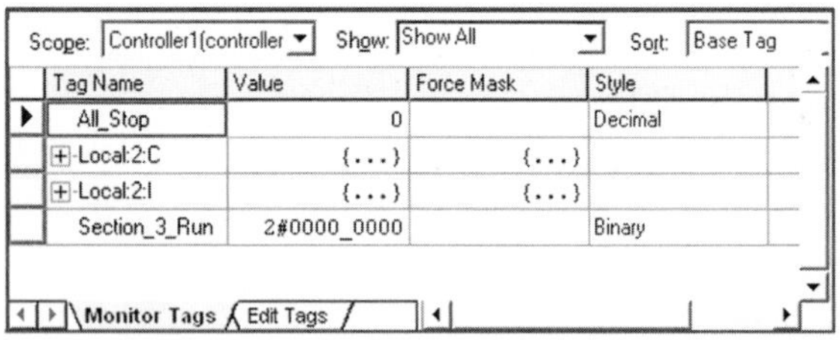

Figure 15-4 Tag window for question 28.

28. Figure 15-4 shows as an example of the ____ tag window.   28._____

a) new   c) edit

b) monitor   d) define

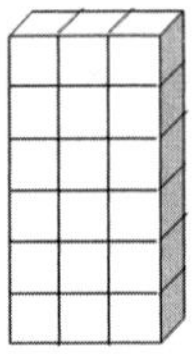

Figure 15-5 Array for question 29.

29. Figure 15-5 shows as an example of a ____ dimensional array.   29._____
a) 1   c) 3
b) 2   d) 4

30. An array type tag can   30._____
a) hold only one type of data.   c) hold up to 100 values.
b) hold more than one type of data.   d) only be used in the main program.

TEST 2

Place the answers to the following questions in the answer column at the right.

1. The internal memory organization of a ControlLogix (CLX) controller is fixed and automatically configured when beginning a project. (True or False) 1._____

2. RSLinx software is used to set up a communications link between the RSLogix 5000 (a) ____ software and the ControlLogix (b) ____. 2a._____ 2b._____

3. CLX modules will not work unless they have been properly configured. (True or False) 3._____

4. A CLX controller can hold and execute several projects at a time. (True or False) 4._____

5. A project is the overall complete application. (True or False) 5._____

6. Each folder of the controller organizer tree groups common functions together. (True or False) 6._____

7. A task contains an executable code. (True or False) 7._____

8. The three types of task execution are (a) ____, (b) ____, and (c) ____. 8a._____ 8b._____ 8c._____

9. More than one task can execute at a time. (True or False) 9._____

10. An application can be broken into a number of tasks. (True or False) 10._____

11. Periodic tasks function as timed interrupts. (True or False) 11._____

12. Only one task may be executing at any given time. (True or False) 12._____

13. The lowest priority task execution is ____. 13._____

14. A continuous task executes any time a periodic or event-based task is not executing. (True or False)

14._____

15. Programs execute in the order in which they are displayed in the controller organizer under their ____.

15._____

16. Unscheduled programs can be downloaded to the controller but remain unscheduled until needed. (True or False)

16._____

17. Logic is written in routines. (True or False)

17._____

18. A routine in CLX is similar to the program in most other PLCs. (True or False)

18._____

19. It is possible to use different programming languages within any one routine. (True or False)

19._____

20. ControlLogix controllers use a ____ based addressing structure.

20._____

21. ____ refers to which programs have access to a tag.

21._____

22. Program scoped tags are available to all programs in a project. (True or False)

22._____

23. A tag is a meaningful name for a memory location. (True or False)

23._____

24. The two scopes for tags in CLX controllers are (a) ____ scope and (b) ____ scope.

24a._____
24b._____

25. A ____ scope tag consists of data that are accessible by all routines within a controller.

25._____

26. The scope of a tag must be declared when you create the tag. (True or False)

26._____

27. I/O tags are automatically created as ____ scope tags.

27._____

28. Controller scoped tags consist of data that can only be accessed by the routine within a single program. (True or False) 28._____

29. BOOL type tags can be used to hold binary numbers up to 16 digits long. (True or False) 29._____

30. DINT type tags can be used to hold binary integers (True or False) 30._____

31. A structure-type tag can only hold data of the same type. (True or False) 31._____

32. The CLX timer instruction is an example of a predefined structure. (True or False) 32._____

33. When you add I/O modules to a project, a number of defined tags are automatically created. (True or False) 33._____

34. An array occupies a continuous block of multiple pieces of data. (True or False) 34._____

35. Arrays are similar to tables of values. (True or False) 35._____

36. An array can hold multiple types of data. (True or False) 36._____

37. A array can have up to three dimensions. (True or False) 37._____

Part 2 Bit Level Programming

TEST 1

Choose the letter that best completes the statement.

1. During each program scan the processor   1._____
a) reads all inputs and takes these values to control the outputs according to the program.
b) reads all outputs and takes these values to control the inputs according to the program.
c) reads all inputs and adjusts these values to control the outputs according to the program.
d) reads the program and adjusts all inputs and outputs accordingly.

2. CLX bit level instructions require _____ addresses.   2._____
a) DINT   c) BOOL
b) INT   d) REAL

3. When creating a ladder rung   3._____
a) all input instructions must be to the right of an output instruction.
b) a rung must contain at least one input instruction.
c) the last instruction must be an output instruction.
d) all of these

4. The XIC instruction   4._____
a) processor checks for a logic input of 0 or 1.
b) returns a true value if the input is logic 1.
c) returns a true value if the input is 0.
d) both a and b

5. The XIO instruction   5._____
a) processor checks for a logic input of 0 or 1.
b) returns a true value if the input is logic 1.
c) returns a true value if the input is 0.
d) both a and c

6. The OTE instruction

a) is true when the rung associated with it has logic continuity.

b) when true can be used to energize an output.

c) when true can be used to set a value in memory to 1.

d) all of these

6._____

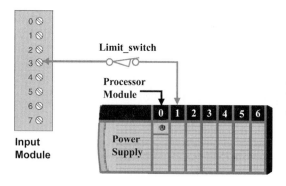

Figure 15-6 Tag-based address for question 7.

7. The physical address for the tag Limit_switch of Figure 15-6 would be

7._____

a) Local:I:3. Data.1.

b) Local:I:1. Data.3.

c) Local:O:3. Data.1.

d) Local:O:1. Data.3.

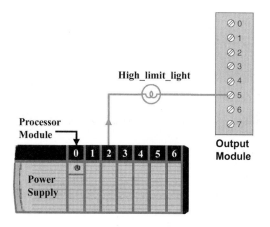

Figure 15-7 Tag-based address for question 8.

8. The physical address for the tag High_limit_light of Figure 15-7 would be

8._____

a) Local:I:2. Data.5.

b) Local:I:5. Data.2.

c) Local:O:2. Data.5.

d) Local:O:5. Data.2.

9. The real-world pushbutton associated with an XIC stop contact in a motor start/stop control logic is

9._____

a) normally closed.

b) normally open.

c) timed closed.

d) timed open.

10. The real-world pushbutton associated with an XIC start contact 10._____
in a motor start/stop control logic is

a) normally closed.   c) timed closed.

b) normally open.   d) timed open.

Figure 15-8 Window for question 11.

11. Figure 15-8 is an example of creating a tag in the 11._____

a) New Tag window.   c) Edit Tag window.

b) Controller Organizer window.   d) Monitor Tag window.

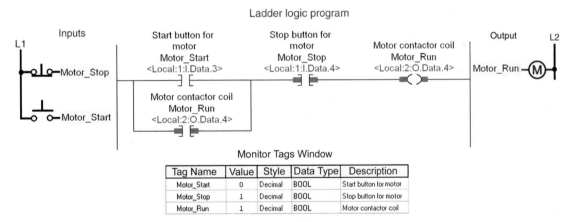

Figure 15-9 Start/Stop program for question 12.

12. For the motor start/stop program of Figure 15-9, what would the value 12._____
of the tags be when the motor is not operating?

a) Motor_Start (0), Motor_Stop (1), Motor_Run (0)

b) Motor_Start (0), Motor_Stop (0), Motor_Run (0)

c) Motor_Start (1), Motor_Stop (1), Motor_Run (0)

d) Motor_Start (1), Motor_Stop (1), Motor_Run (1)

13. Internal relay instructions are used when

a) real-world input field devices are needed as inputs.

b) real-world output field devices are needed as outputs.

c) real-world field devices are not needed as input or output reference instructions.

d) both a and b

13._____

14. The OTL instruction

a) is a retentive output instruction.

b) once on will stay on even if the status of its input becomes false.

c) is used in conjunction with an OTU instruction with the same referenced tag.

d) all of these

14._____

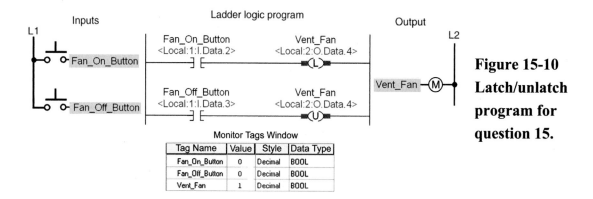

Figure 15-10 Latch/unlatch program for question 15.

15. For the latch/unlatch program of Figure 15-10, what would the value of the tags be after momentary actuation of the Fan_Off_Button?

15._____

a) Fan_On_Button (0), Fan_Off_Button (1), Vent_Fan (0)

b) Fan_On_Button (0), Fan_Off_Button (0), Vent_Fan (0)

c) Fan_On_Button (1), Fan_Off_Button (1), Vent_Fan (0)

d) Fan_On_Button (0), Fan_Off_Button (1), Vent_Fan (0)

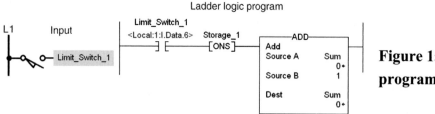

Figure 15-11 One shot program for question 16.

16. For the one shot program of Figure 15-11, the ADD function is executed   16._____
   a) only once per actuation of the limit switch.
   b) as long as the limit switch is closed
   c) as long as the limit switch is opened.
   d) either b or c

TEST 2

Place the answers to the following questions in the answer column at the right.

1. A CLX controller executes the program in real time. (True or False)

1._____

2. A CLX processor can update the input tag from the field and write the output tag to the field at different points during execution of the program. (True or False)

2._____

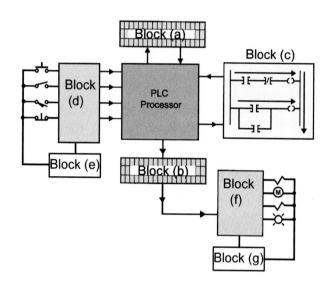

Figure 15-12 Block diagram for question 3.

3. For the Logix controller block diagram shown in Figure 15-12, block (a) is ____, block (b) is ____, block (c) is ____, block (d) is ____, block (e) is ____, block (f) is ____, and block (g) is ____.

3a._____

3b._____

3c._____

3d._____

3e._____

3f._____

3g._____

4. A rung does not need to contain any input instructions. (True or False)

4._____

5. Contacts are output instructions. (True or False)

5._____

6. Coils are input instructions. (True or False)

6._____

7. XIO is an acronym for examine if ____.

7._____

8. XIC is an acronym for examine if ____.

8._____

9. OTE is an acronym for output ____.

9._____

10. Output instructions can be placed in series on a rung in CLX logic. (True or False)

10._____

11. Output instructions can be placed in parallel on a rung in CLX logic. (True or False)

11._____

12. A tag is a text-based name for an area of the controller where ____ is stored.

12._____

13. Tag-based addressing is not tied to specific memory locations in the memory structure. (True or False)

13._____

14. The ____ tags window shows the state of the tags created for a program.

14._____

15. Internal relay instructions are used when real-world field devices are needed as input or output reference instructions. (True or False)

15._____

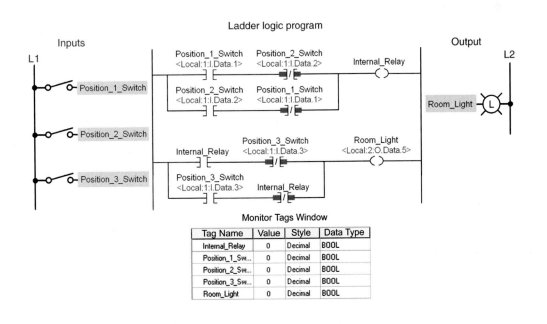

Figure 15-13 Internal relay program for question 16.

16. For the internal relay program shown in Figure 15-13,
when Position_1_Switch is closed and the other two switches are open,
what value is stored in each of the following?
Internal_Relay (a), Position_1_Switch (b), Position_2_Switch (c),
Position_3_Switch (d), and Room_Light (e)

16a._____

16b._____

16c._____

16d._____

16e._____

17. OTU is an acronym for the output ____ instruction.

17._____

18. OTL is an acronym for the output ____ instruction.

18._____

19. ONS is an acronym for the one ____ instruction.

19._____

20. An ONS instruction can be used to turn an output on for one scan.

20._____

Programming Assignments for Part 2

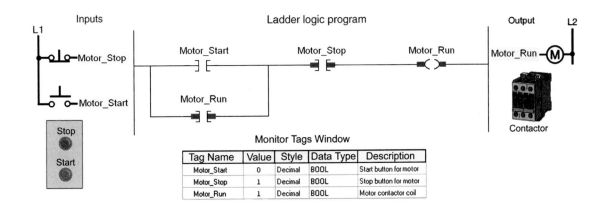

Figure 15-14 Start/stop motor control program for assignment 1.

1) The start/stop motor control program of Figure 15-14 is described in the text.
a) Prepare an I/O connection diagram and ladder logic program that will simulate its operation. Utilize field devices found on your ControlLogix installation. Enter the program into the controller, and monitor its operation.
b) Modify the operation to include a second start and stop pushbutton station.

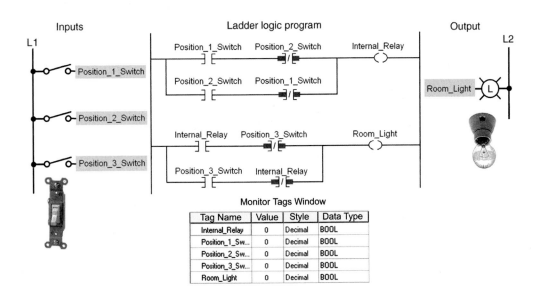

Figure 15-15 Internal relay program for assignment 2.

2) The internal relay program of Figure 15-15 is described in the text.

a) Prepare an I/O connection diagram and ladder logic program that will simulate its operation. Utilize field devices found on your ControlLogix installation. Enter the program into the controller, and monitor its operation.

b) Modify the operation to include an additional single pole switch that will implement ON/OFF control of the light from a fourth position.

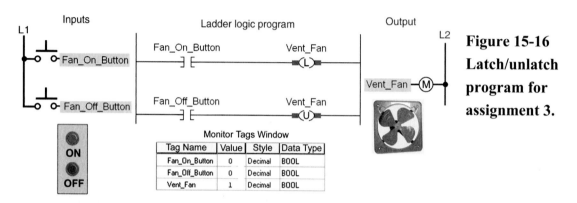

Figure 15-16 Latch/unlatch program for assignment 3.

3) The latch/unlatch program of Figure 15-16 is described in the text.

a) Prepare an I/O connection diagram and ladder logic program that will simulate its operation. Utilize field devices found on your ControlLogix installation. Enter the program into the controller, and monitor its operation.

b) Modify the operation to include an additional on/off pushbutton station.

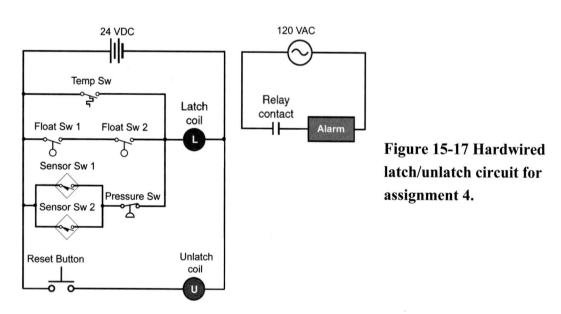

Figure 15-17 Hardwired latch/unlatch circuit for assignment 4.

4) Prepare an I/O connection diagram and ControlLogix program that will simulate the operation of the hardwired latching relay alarm circuit shown in Figure 15-17 and described in the text. Utilize field devices found on your ControlLogix installation. Enter the program into the controller, and monitor its operation.

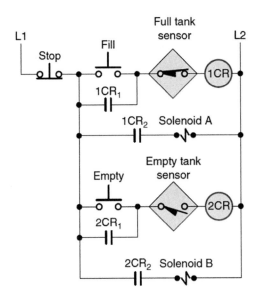

Figure 15-18 Hardwired tank filling and emptying operation for assignment 5.

5) Prepare an I/O connection diagram and ControlLogix program that will simulate the operation of the hardwired tank filling and emptying operation shown in Figure 15-18 and described in the text. Utilize field devices found on your ControlLogix installation. Enter the program into the controller, and monitor its operation.

Part 3   Programming Timers

TEST 1

Choose the letter that best completes the statement.

1. Timers are used to   1._____
a) turn outputs on and off after a time delay.
b) turn outputs on or off for a set amount of time.
c) keep track of the time an output is on or off.
d) all of these

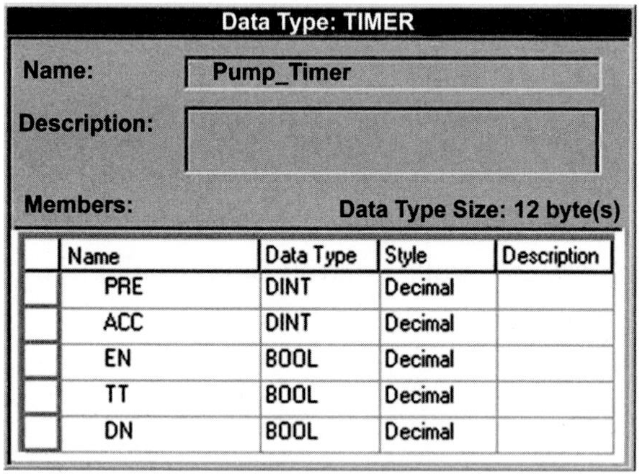

Data Type: TIMER			
Name: Pump_Timer			
Description:			
Members: Data Type Size: 12 byte(s)			
Name	Data Type	Style	Description
PRE	DINT	Decimal	
ACC	DINT	Decimal	
EN	BOOL	Decimal	
TT	BOOL	Decimal	
DN	BOOL	Decimal	

Figure 15-19 Timer structure for question 2.

2-1. For the timer structure shown in Figure 15-19, which member specifies   2-1._____
the value the timer must accumulate to reach the desired time delay?
a) PRE   c) ACC
b) EN   d) TT

2-2. Which member indicates that accumulated value   2-2._____
is equal to the preset value?
a) PRE   c) ACC
b) EN   d) TT

2-3. Which member indicates that a timing operation   2-3._____
is in process?
a) PRE   c) ACC
b) EN   d) TT

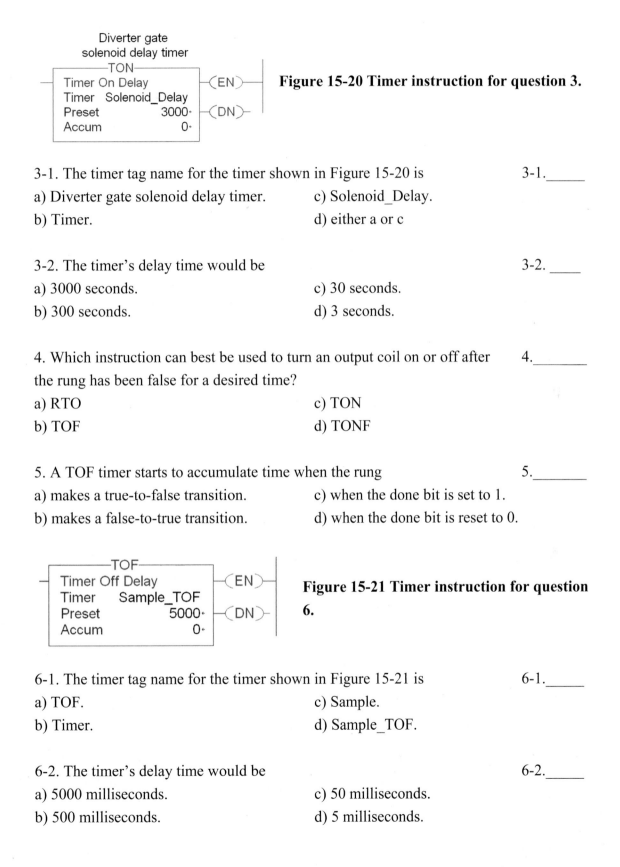

Figure 15-20 Timer instruction for question 3.

3-1. The timer tag name for the timer shown in Figure 15-20 is

3-1._____

a) Diverter gate solenoid delay timer.   c) Solenoid_Delay.

b) Timer.   d) either a or c

3-2. The timer's delay time would be

3-2._____

a) 3000 seconds.   c) 30 seconds.

b) 300 seconds.   d) 3 seconds.

4. Which instruction can best be used to turn an output coil on or off after the rung has been false for a desired time?

4._____

a) RTO   c) TON

b) TOF   d) TONF

5. A TOF timer starts to accumulate time when the rung

5._____

a) makes a true-to-false transition.   c) when the done bit is set to 1.

b) makes a false-to-true transition.   d) when the done bit is reset to 0.

Figure 15-21 Timer instruction for question 6.

6-1. The timer tag name for the timer shown in Figure 15-21 is

6-1._____

a) TOF.   c) Sample.

b) Timer.   d) Sample_TOF.

6-2. The timer's delay time would be

6-2._____

a) 5000 milliseconds.   c) 50 milliseconds.

b) 500 milliseconds.   d) 5 milliseconds.

7. A RTO retentive on-delay timer retains its ACC value even if

7._____

a) the rung goes false.

b) the processor is placed in the program mode.

c) power to the processor is temporarily interrupted.

d) all of these

8. The RTO timer's ACC value is reset to zero by a

8._____

a) RES instruction with a different tag name.

b) RES instruction with the same tag name.

c) true-to-false transition of the timer rung.

d) false-to-true transition of the timer rung.

TEST 2

Place the answers to the following questions in the answer column at the right.

1. The timer address in the ControlLogix controller is a predefined _____ of   1._____
the TIMER data type.

2. The on-delay timer (TON) is a retentive timer. (True or False)   2._____

3. The timer's delay time would equal the value in the ACC multiplied by   3._____
the time base. (True or False)

4. Retentive timers lose the accumulated time every time the rung condition   4._____
becomes false. (True or False)

5. The time increment used in CLX timers is milliseconds. (True or False)   5._____

6. The TON instruction produces a(n) _____-delay timer.   6._____

7. A TON timer begins accumulating time when the rung conditions   7._____
become true. (True or False)

8. The timer off-delay (TOF) instruction can be used to turn an output coil   8._____
on or off after the rung has been false for a desired time. (True or False)

9. A TOF timer begins accumulating time when the rung conditions   9._____
become true. (True or False)

10. The _____ instruction is used to reset a timer's accumulated value.   10._____

Programming Assignments for Part 3

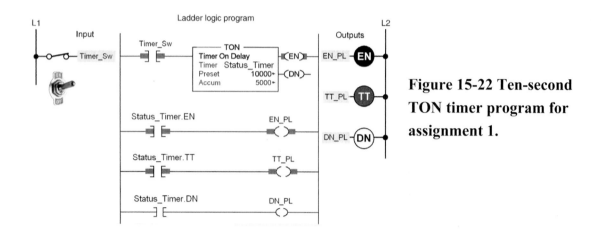

Figure 15-22 Ten-second TON timer program for assignment 1.

1) The ten-second TON timer program of Figure 15-22 is described in the text.
a) Prepare an I/O connection diagram and ladder logic program that will simulate its operation. Utilize field devices found on your ControlLogix installation. Enter the program into the controller, and monitor its operation.
b) Modify the program with an additional rung added that will energize a solenoid whenever the timer is enabled and timing.

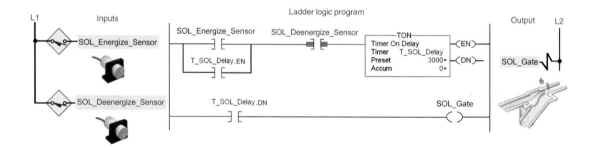

Figure 15-23 TON timer program used to delay the operation of a diverter for assignment 2.

2) A TON timer program used to delay the operation of a diverter gate is shown in Figure 15-23 and described in the text. Prepare an I/O connection diagram and ladder logic program that will simulate its operation. Utilize field devices found on your ControlLogix installation. Enter the program into the controller and monitor its operation.

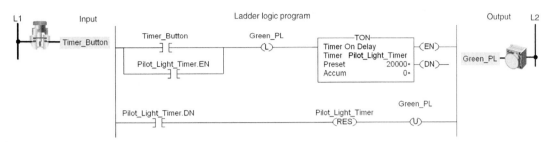

Figure 15-24 Pilot light TON timer program for assignment 3.

3) The pilot light TON timer program of Figure 15-24 is described in the text. Prepare an I/O connection diagram and ladder logic program that will simulate its operation. Utilize field devices found on your ControlLogix installation. Enter the program into the controller, and monitor its operation.

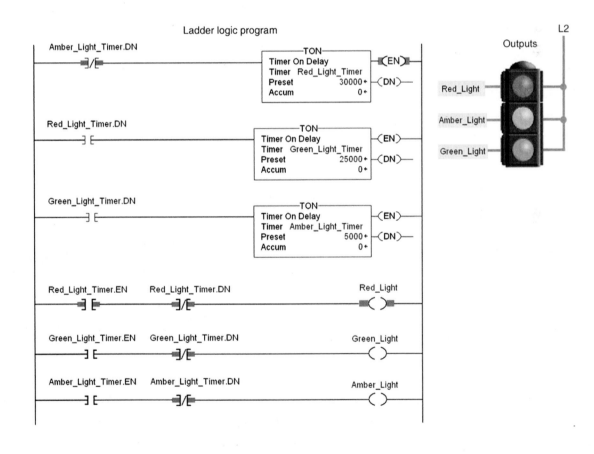

Figure 15-25 Traffic control program for assignment 4.

4) The traffic control program of Figure 15-25 is described in the text.

a) Prepare an I/O connection diagram and ladder logic program that will simulate its operation. Utilize field devices found on your ControlLogix installation. Enter the program into the controller, and monitor its operation.

b) Modify the program to extend the green light ON time to 40 seconds.

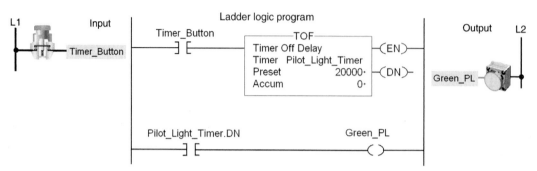

Figure 15-26 Pilot light TOF timer program for assignment 5.

5) The pilot light TOF timer program of Figure 15-26 is described in the text. Prepare an I/O connection diagram and ladder logic program that will simulate its operation. Utilize field devices found on your ControlLogix installation. Enter the program into the controller, and monitor its operation.

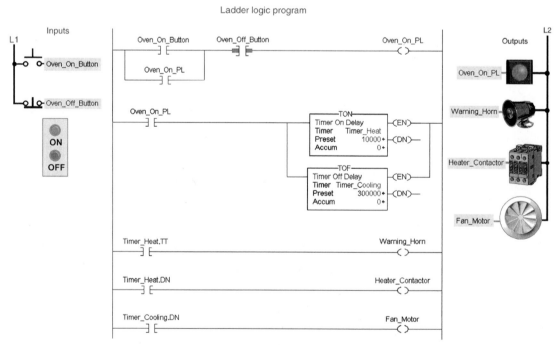

Figure 15-27 Timer control of a heating process program for assignment 6.

6) The timer control of a heating process program of Figure 15-27 is described in the text. Prepare an I/O connection diagram and ladder logic program that will simulate its operation. Utilize field devices found on your ControlLogix installation. Enter the program into the controller, and monitor its operation.

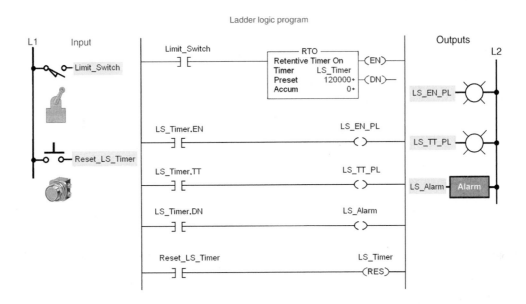

Figure 15-28 Limit switch RTO timer program for assignment 7.

7) The limit switch RTO timer program of Figure 15-28 is described in the text.
 a) Prepare an I/O connection diagram and ladder logic program that will simulate its operation. Utilize field devices found on your ControlLogix installation. Enter the program into the controller, and monitor its operation.
 b) Modify the program to include a warning light that comes on with the alarm to signal that the timer has timed out.

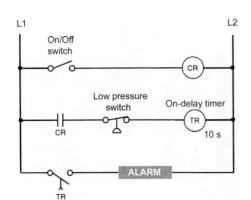

Figure 15-29 Hardwired TON alarm circuit for assignment 8.

8) Prepare an I/O connection diagram and ControlLogix program that will simulate the operation of the hardwired TON alarm circuit shown in Figure 15-29 and described in the text. Utilize field devices found on your ControlLogix installation. Enter the program into the controller, and monitor its operation.

Part 4   Programming Counters

TEST 1

Choose the letter that best completes the statement.

1. Counters count   1._____
a) true-to-false transitions of the counter rung.
b) false-to-true transitions of the counter rung.
c) the length of time a counter rung is true.
d) the length of time a counter rung is false.

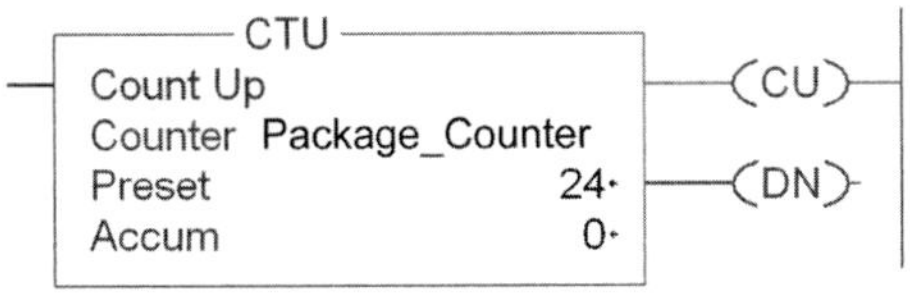

Figure 15-30 Counter instruction for question 2.

2-1. For the counter instruction shown in Figure 15-30, which parameter   2-1._____
specifies the value the counter must reach before the done bit turns on?
a) Count Up   c) Preset
b) Package_Counter   d) Accum

2-2. Which parameter indicates the number of transitions of the counter rung? 2-2._____
a) Count Up   c) Preset
b) Package_Counter   d) Accum

2-3 What is the counter's tag name?   2-3._____
a) Count Up   c) Preset
b) Package_Counter   d) Accum

3. The counter Accum value is reset to zero by a   3._____
a) RES instruction with a different tag name.
b) RES instruction with the same tag name.
c) true-to-false transition of the counter rung.
d) false-to-true transition of the counter rung.

4. A CTD counter will cause the accumulated value to _____ when there is a false-to-true transition of the counter rung.

a) reset to zero
c) decrease by one
b) increase by one
d) remain the same

4._____

5. The ControlLogix CTD instruction is typically used with a CTU instruction that references

a) the same counter structure.
c) the same data table.
b) a different counter structure.
d) a different data table.

5._____

TEST 2

Place the answers to the following questions in the answer column at the right.

1. A counter counts the change of state of an external trigger signal. (True or False)

1._____

2. The two basic counter types are the (a) ____-counter and the (b) ____-counter.

2a._____
2b._____

3. All counters are nonretentive. (True or False)

3._____

4. The acronym CTD stands for a count-____ counter.

4._____

5. The acronym CTU stands for a count-____ counter.

5._____

6. A counter retains its Accum value if the rung goes false. (True or False)

6._____

Programming Assignments for Part 4

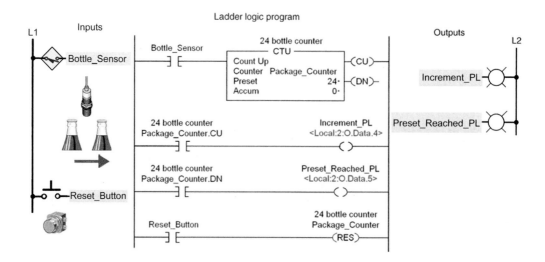

Figure 15-31 Count-up counter program for assignment 1.

1) The count-up counter program of Figure 15-31 is used to count packets of bottles and is described in the text.

a) Prepare an I/O connection diagram and ladder logic program that will simulate its operation. Utilize field devices found on your ControlLogix installation. Enter the program into the controller, and monitor its operation.

b) Modify the program to count 6 bottle packets.

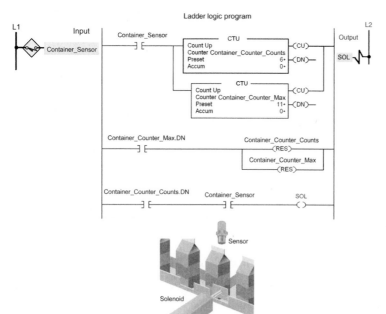

Figure 15-32 Count-up counter program for assignment 2.

2) The count-up counter program of Figure 15-32 is used to remove containers from a conveyor line and is described in the text. Prepare an I/O connection diagram and ladder logic program that will simulate its operation. Utilize field devices found on your ControlLogix installation. Enter the program into the controller, and monitor its operation.

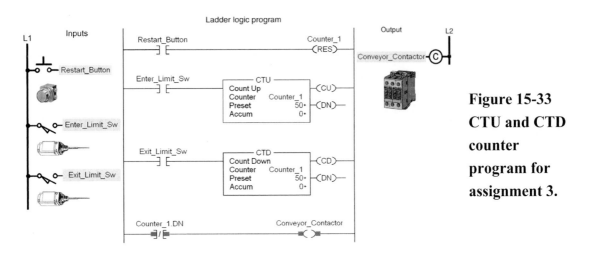

Figure 15-33 CTU and CTD counter program for assignment 3.

3) An example of CTU and CTD counters used together is shown in Figure 15-33 and described in the text.

 a) Prepare an I/O connection diagram and ladder logic program that will simulate its operation. Utilize field devices found on your ControlLogix installation. Enter the program into the controller, and monitor its operation.

 b) Modify the program to include a red pilot light to indicate entry of a part into the buffer zone and a green pilot light to indicate exit of a part from the buffer zone.

4) Write a ControlLogix program, complete with tags, for an up/down counter used to keep track of cars entering and exiting a parking lot. The program requirements for this application are summarized as follows:

- The parking lot holds 30 vehicles.
- There is an entrance vehicle sensor and an exit vehicle sensor.
- When the parking lot is full, a Lot Full sign is illuminated.
- Whenever a car exits the lot, a Caution Buzzer/Light is activated to warn pedestrians.

Utilize field devices found on your ControlLogix installation. Enter the program into the controller, and monitor its operation.

Part 5 Math, Comparison, and Move Instructions

TEST 1

Choose the letter that best completes the statement.

1. The ADD instruction adds values from   1._____
a) Source A and Source B and stores the result in Source C.
b) Source B and Source C and stores the result in the Source A.
c) Source A and Source B and stores the result in the Dest.
d) Source A and Dest and stores the result in the Dest.

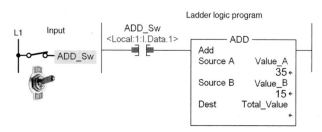

Figure 15-34 ADD instruction rung for question 2.

2. For the ADD instruction rung shown in Figure 15-34, the value stored   2._____
in Dest would be
a) 1.   c) 50.
b) 35.   d) 15.

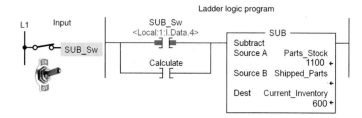

Figure 15-35 SUB instruction rung for question 3.

3. For the SUB instruction rung shown in Figure 15-35, the number of   3._____
shipped parts would be
a) 100.   c) 600.
b) 500.   d) 1700.

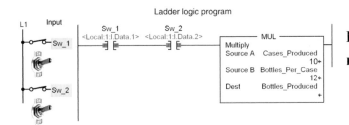

Figure 15-36 MUL instruction rung for question 4.

4. For the MUL instruction rung shown in Figure 15-36, the number of bottles produced would be   4._____
a) 120.   c) 200.
b) 1200.   d) 800.

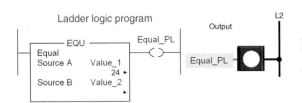

Figure 15-37 EQU instruction rung for question 5.

5. For the EQU instruction rung shown in Figure 15-37, what value(s)   5._____
stored in Source B would make the instruction logically true?
a) 24   c) 0 to 23
b) 25   d) 0 to 24

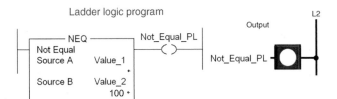

Figure 15-38 NEQ instruction rung for question 6.

6. For the NEQ instruction rung shown in Figure 15-38, what value stored   6._____
in Source A would make the instruction logically true?
a) 10   c) 146
b) 98   d) All of these

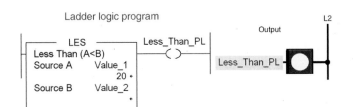

Figure 15-39 LES instruction rung for question 7.

7. For the LES instruction rung shown in Figure 15-39, what Value_2 would   7._____
make the instruction logically true?

a) 5   c) 25

b) 15   d) All of these

Ladder logic program

GRT
Greater Than (A>B)
Source A   Value_1
Source B   Value_2
               300

Greater_Than_PL
( )

Output
Greater_Than_PL

L2

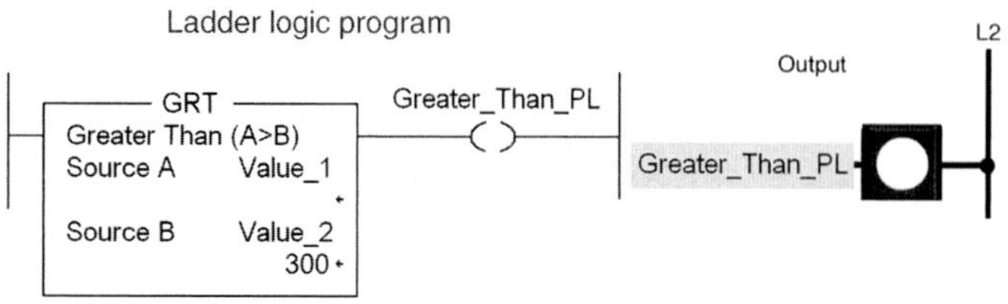

Figure 15-40 GRT instruction rung for question 8.

8. For the GRT instruction rung shown in Figure 15-40, what Value_1   8._____
would make the instruction logically true?

a) 301   c) 299

b) 300   d) All of these

TEST 2

Place the answers to the following questions in the answer column at the right.

1. Source is a value that is input to a math instruction. (True or False)   1._____

2. Dest (destination) is where the ____ of the math instruction is stored.   2._____

3. Math instructions always send a REAL number result to the destination   3._____
tag. (True or False)

4. Math instructions always send a DIN number result to the destination tag.   4._____
(True or False)

5. Math instructions always send a number type that matches the type   5._____
of the destination tag. (True or False)

6. Compare instructions are used to compare two ____.   6._____

7. CMP instructions allow the programmer to enter complex expressions   7._____
in one instruction. (True or False)

8. The MOV instruction can move the contents of one memory location   8._____
to another location. (True or False)

Programming Assignments for Part 5

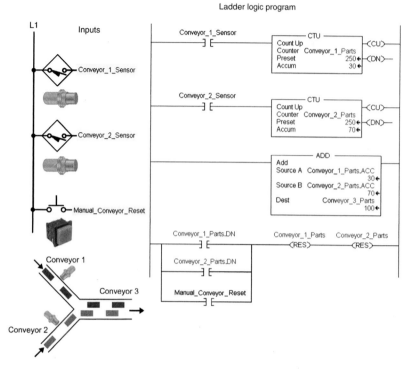

Figure 15-41 Parts tracking program for assignment 1.

1) The program of Figure 15-41 is used as part of a parts tracking system and is described in the text. Prepare an I/O connection diagram and ladder logic program that will simulate its operation. Utilize field devices found on your ControlLogix installation. Enter the program into the controller, and monitor its operation.

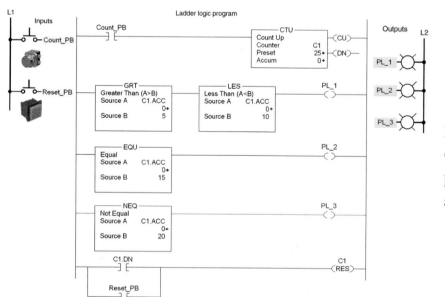

Figure 15-42 Comparison program for assignment 2.

2) The comparison program of Figure 15-42 is used to test the accumulated value of a counter and is described in the text. Prepare an I/O connection diagram and ladder logic program that will simulate its operation. Utilize field devices found on your ControlLogix installation. Enter the program into the controller, and monitor its operation.

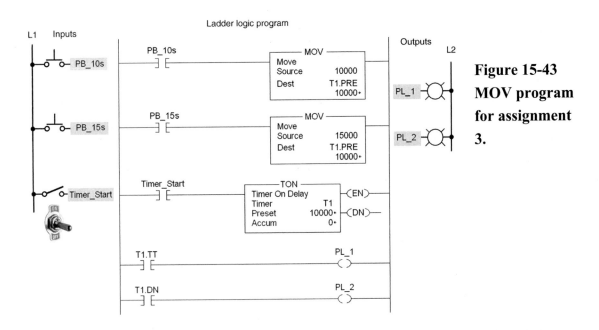

Figure 15-43 MOV program for assignment 3.

3) The MOV program of Figure 15-43 is used to create a variable preset timer and is described in the text. Prepare an I/O connection diagram and ladder logic program that will simulate its operation. Utilize field devices found on your ControlLogix installation. Enter the program into the controller, and monitor its operation.

Part 6   Function Block Programming

TEST 1

Choose the letter that best completes the statement.

1. A functional block diagram (FBD) is a _____ programming language.   1._____

a) textual   c) graphical

b) contact and coil   d) schematic

2. A sheet of an FBD consists of   2._____

a) function blocks joined together with wires.

b) a series of input and output blocks.

c) gate symbols with connecting input and output wires.

d) interconnected series of blocks containing circuit schematics.

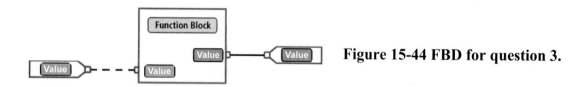   **Figure 15-44 FBD for question 3.**

3-1. For the FBD of Figure 15-44, the solid line indicate: 3-1._____

a) no type of data is present.

b) a Boolean value is present.

c) an integer or real value is present.

d) a combination of data types are present.

3-2. The dash line indicates   3-2._____

a) no type of data is present.

b) a Boolean value is present.

c) an integer or real value is present.

d) a combination of data types are present.

346

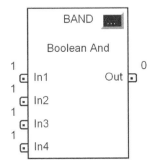

Figure 15-45 BAND function block for question 4.

4. For the BAND function block of Figure 15-45, the   4._____

a) tag name for the block is placed above it.

b) 1 and 0 next to the inputs and output identify their logical state.

c) dots on the pins indicate BOOL type data are required.

d) all of these

5. Which element of a function block diagram represents a value from an   5._____
input device that brings a value into a function block?

a) Wire   c) Output Reference

b) Function Block   d) Input Reference

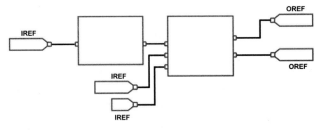

Figure 15-46 Function block diagram for question 6.

6-1. The IREFs shown in Figure 15-46   6-1._____

a) are used to send a value to an output device or tag.

b) are used to receive a value from an input device or tag.

c) must be contain tags.

d) both b and c

6-2. The OREFs shown   6- 2._____

a) are used to send a value to an output device or tag.

b) are used to receive a value from an input device or tag.

c) must be contain tags.

d) both a and c

7. ICONs and OCONs 7._____

a) are used to exchange information between function blocks.

b) require a unique tag name.

c) must have the same tag name.

d) all of these

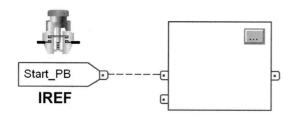

Figure 15-47 IREF for question 8.

Start_PB

IREF

8. With reference to Figure 15-47, if the pushbutton is momentarily 8._____ actuated, the data in the IREF

a) change immediately. c) execute immediately.

b) are latched in for one scan. d) execute after a delay of one scan.

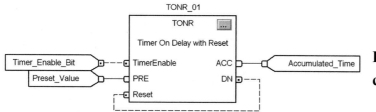

Figure 15-48 Wire connection for question 9.

9. With reference to Figure 15-48, the wire connected between the DN 9._____ and Reset pins

a) represents a DINT signal path. c) establishes data latching.

b) creates a feedback loop. d) all of these

TEST 2

Place the answers to the following questions in the answer column at the right.

1. Similar to ladder logic programming, functional block diagram (FBD) programming uses power rails. (True or False)

1._____

2. The workplace of an FBD is known as a(n) ____.

2._____

3. The four basic elements of a FBD are (a) ____, (b) ____, (c) ____, and (d) ____.

3a._____
3b._____
3c._____
3d._____

4. Function blocks contain nonexecutable code. (True or False)

4._____

5. Add-on instructions are special purpose instructions that can be purchased and added to the instruction set. (True or False)

5._____

6. The acronym IREF stands for ____ reference.

6._____

7. The acronym OREF stands for ____ reference.

7._____

8. The acronym ICON stands for input ____.

8._____

9. Function block connections are made using wires and pins. (True or False)

9._____

10. The pins on the left of a function block are output pins, and those on the right are input pins. (True or False)

10._____

11. Wire ____ are used to create a signal path without using a wire.

11._____

12. Each OCON must have at least one corresponding ICON. (True or False)

12._____

Programming Assignments for Part 6

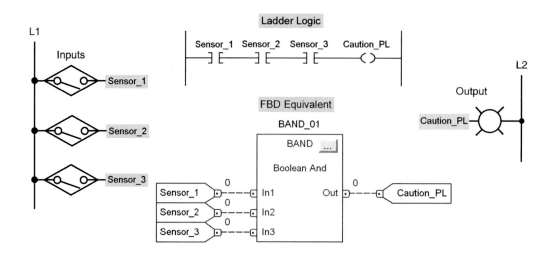

Figure 15-49 AND logic program for assignment 1.

1) The programs of Figure 15-49 show the ladder logic and the FBD equivalent for a three-input AND logic. Prepare a function block program that will simulate its operation. Utilize field devices found on your ControlLogix installation. Enter the program into the controller, and verify its operation.

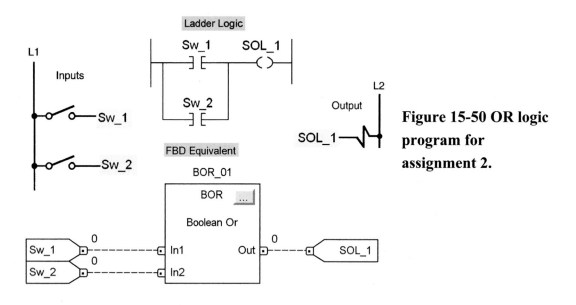

Figure 15-50 OR logic program for assignment 2.

2) The programs of Figure 15-50 show the ladder logic and the FBD equivalent for a two-input OR logic. Prepare a function block program that will simulate its operation. Utilize field devices found on your ControlLogix installation. Enter the program into the controller, and verify its operation.

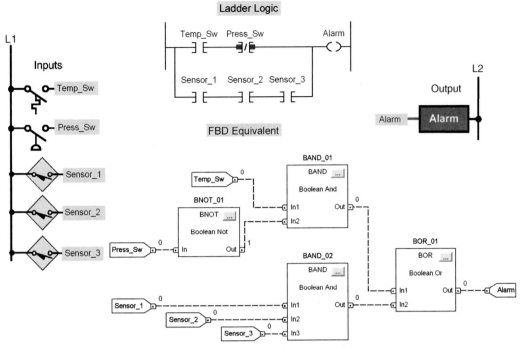

Figure 15-51 Multiple inputs program for assignment 3.

3) The programs of Figure 15-51 show the ladder logic and the FBD equivalent for a combination of multiple inputs logic. Prepare a function block program that will simulate its operation. Utilize field devices found on your ControlLogix installation. Enter the program into the controller, and verify its operation.

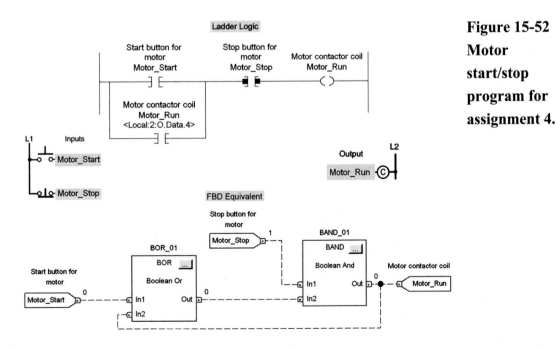

Figure 15-52 Motor start/stop program for assignment 4.

4) The programs of Figure 15-52 show the ladder logic and the FBD equivalent for motor start/stop control logic. The operation of the FBD program is explained in the text.

a) Prepare a function block program that will simulate its operation. Utilize field devices found on your ControlLogix installation. Enter the program into the controller, and verify its operation.

b) Modify the FBD program to include a second start/stop pushbutton station.

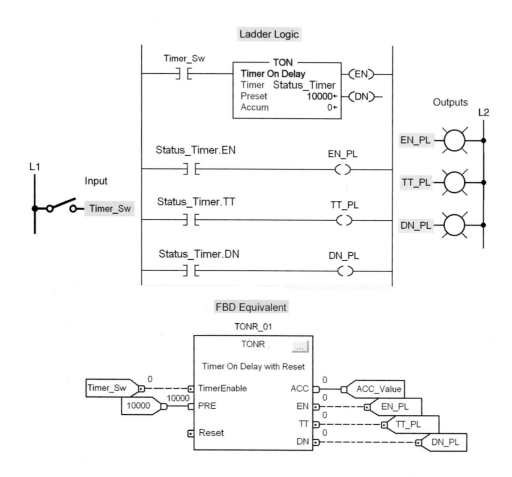

Figure 15-53 On-delay timer program for assignment 5.

5) The programs of Figure 15-53 show the ladder logic and the FBD equivalent for an on-delay timer. The operation of the FBD program is explained in the text. Prepare a function block program that will simulate its operation. Utilize field devices found on your ControlLogix installation. Enter the program into the controller, and verify its operation.

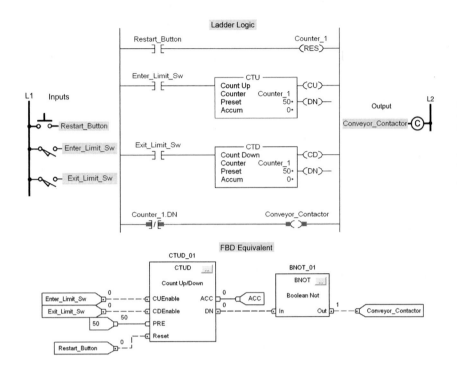

Figure 15-54
Up/down counter
program for
assignment 6.

6) The programs of Figure 15-54 show the ladder logic and the FBD equivalent for an up/down counter program. The operation of the FBD program is explained in the text.

a) Prepare a function block program that will simulate its operation. Utilize field devices found on your ControlLogix installation. Enter the program into the controller, and verify its operation.

b) Modify the FBD program to include the following three pilot lights:

- PL_1 to come on when a part enters.
- PL_2 to come on when a part exits.
- PL_3 to come on when the buffer zone is full.

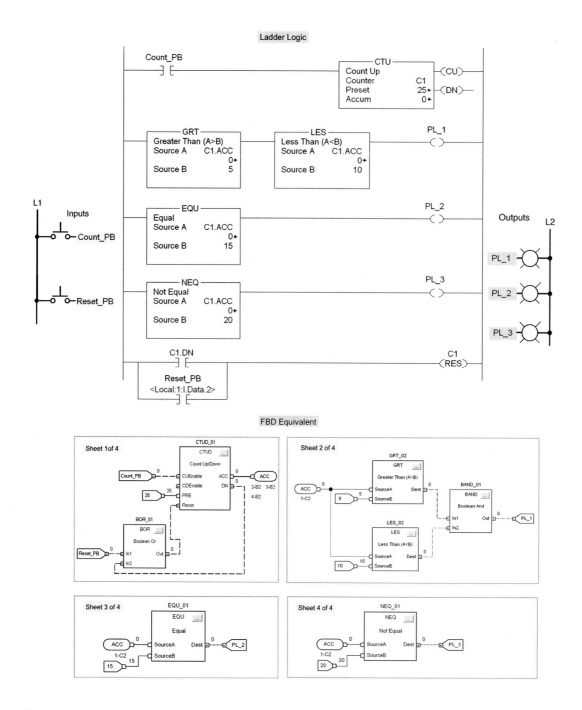

Figure 15-55 Counter program for assignment 7.

7) The programs of Figure 15-55 show the ladder logic and the FBD equivalent for a program used to test the accumulated value of a counter. The operation of the FBD program is explained in the text.

a) Prepare a function block program that uses 4 sheets to simulate its operation. Utilize field devices found on your ControlLogix installation. Enter the program into the controller, and verify its operation.

b) Modify the FBD program to include the following three pilot lights:

- PL_1 to be on for an accumulated count between 0 and 5
- PL_2 to be on for an accumulated count of 12.
- PL_3 to be on at all times except when the accumulated count is 15.

8) Write an FBD program that will cause the output, solenoid SOL_1, to be energized when pushbuttons PB_1 is open and PB_2 is closed, and either limit switch LS_1 is open or limit switch LS_2 is closed. Assume all pushbuttons and limit switches are of the normally open type. Utilize field devices found on your ControlLogix installation. Enter the program into the controller, and verify its operation.

Notes

Notes

Notes